Behind Every Sunflower

Vortex Based Mathematics,
Sacred Geometry &
The Blueprints of Creation

Gabrielle Ensign

*"The Beauty of Numbers
Behind the Beauty of Nature"*

Published by World Resonance Publishing, LLC
Flagstaff, AZ

Copyright © by Gabrielle Ensign, 2025

All rights reserved.

Published by World Resonance Publishing, LLC
Flagstaff, AZ, 86005

This book may not be reproduced in whole or in part without written permission from the author, except by a reviewer who may quote brief passages in a review; nor may any part of this book be altered and shared electronically without permission from the author. Diagrams created by Gabrielle Ensign may not be altered but may be copied, reproduced and shared, either physically or electronically.

Library of Congress Control Number: 2025901087

ISBN: 979-8-9924399-0-8

Printed in USA by Amazon Kindle Direct Publishing

First edition: January, 2025

"If you want to find the secrets of the universe, think in terms of energy, frequency and vibration."
~Nikola Tesla

Contents

Foreword	9
Introduction	10
Chapter 1	
What is Vortex Based Mathematics?	13
The Fingerprint of Creation	13
Marko Rodin	14
The Doubling Circuit	15
The Rodin Symbol	17
The Polar Pairs	19
Roles to Play	22
The Rodin 147 Shears	23
The Toroid	30
The Gift of Knowing	33
Chapter 2	
The Divine Proportion and Vortex Based Mathematics	35
Phi 1.618	35
The Rodin Phi Diagram	38
Gabby's Rodin Flower	39
Gator's Donut	41
The Root Sum of Phi	45
The Diamond of Fifths	47
Phi is Vortex Based Mathematics	52
Chapter 3	
The Trinity of 3 9 6 and The Divine Proportion	55
The Number 3	55
The Trinity of Creation	56
Holographic Root Sums	58

The Phi Flower of Sums	65
The Rodin Phi Flower of Sums	66
Holographic Sequences	72
The Language of Creation	76

Chapter 4
The Musical Matrix of Creation — 77

All is Frequency	77
Cymatics	77
The Geometry of Planets	78
The Harmonic Numbers and the Circle of Fifths	82
Tuning Frequencies	84
The Musical Notes of Phi?	85
The Family Number Groups	88
Vortex Based Mathematics and the Flower of Life	90
Hertz Relationships	92

Chapter 5
The Music of the Spheres — 95

Pythagoras	95
The Song of Life	95
The Number 12 Throughout History	96
The Number 72 Throughout History	97
The Number 108 Throughout History	97
The Number 432 Throughout History	98
Celestial Harmonics	100
Denying the Truth	102
Ancient Geomathematics	103
All the Same Numbers	106

Chapter 6
The Numbers 1 through 9 and Their Numerical Geometry   107

    The Egg Carton Universe   107

    The 360° Circle   108

    The Numbers 1 and 2   108

    The Numbers 3 and 6   109

    The Numbers 4 and 8   111

    The Number 5   113

    The Number 7   116

    The Decimal of 7   118

    Angle and Ratio   120

    The Number 9   121

    The Enneagram and the Trinity of Creation   125

    Harmonic Alignments   127

Chapter 7
The Number 9 - The Point of Balance in a Bipolar Universe   129

    The Axis of 9   129

    The Enneagram and the Rodin 147 Shears   130

    Holographic Geometry and the Illusionary Universe   133

    The Rodin Trick   135

    The Rodin 147 Shears and the Trinity of Creation   136

    9 Equals Balance, 9 Equals Love   141

Chapter 8
Vortex Based Mathematics and What Comes Next — **145**

- Numerical Synchronicities — 145
- The Ferrocell — 146
- The Hopf Fibration — 147
- Buckminster Fuller and Sphere Packing — 148
- Nassim Haramein and the 64 Tetrahedron Grid — 153
- The Vector Equilibrium — 154
- Spiraling Outward — 155
- Robert Edward Grant and the Prime Wheel — 157
- Vortex Based Mathematics and the Prime Wheel — 159
- The True Planck Length — 161
- The Hebrew Name of God and the Divine Angle — 162
- A Musical Matrix — 164
- Malcolm Bendall and the Thunderstorm Generator — 166
- Bob Greenyer and Ball Lightning — 168
- Charlie Ziese and Universal Phi Scaling — 172
- Vortex Energy Implosion Technologies — 176
- The Rodin Coil Over Unity Transducer — 176
- Sheela Rahman and the XYZ Oscilloscope — 179
- The Keys to a Better World — 182

About the Author — **185**

Image Credits — **186**

Bibliography — **188**

Acknowledgements — **192**

Foreword

Gabby Ensign has dedicated her life to be a bridge between Vortex Based Mathematics and mankind.

Nature is expressing herself with numbers and numbers are alive in everything.

I am very grateful for Gabby's enduring self-sacrifice to bring this discovery to you.

When the invisible connection between numbers and physics is understood then the world will be freed from toil and sin, destruction and death.

This is the bridge between heaven and earth.

-Marko Rodin

Introduction

When you look at a sunflower, what do you see? I see a beautiful yellow flower, but I also see mathematical ratios, geometry, numerical patterns and sequences in harmonic balance, and I see music. I also see a galaxy, a seashell, a beautiful song, and a handshake. I see the work of the mind of creation and the sparkle in your eye. I see the universe and I see life and love.

All this I can see behind what is that sunflower, which is an expression of the life that makes up the entire universe. It works through an energetic dance of movement and form, weaving together all that it takes to create and maintain all that is that we can see. The life that is the universe gives instruction through numerical, mathematical and geometric laws, which is akin to the coded energetic fabric or substance of the universe. It also gives instruction through feelings and intuition.

It is not just a sunflower I see with my physical eyes, it is what is behind the sunflower that I see, which is what I have learned about that sunflower that gives me an understanding in my mind about that sunflower. It lets me see the truth behind that sunflower within this physical space we temporarily occupy, which helps me see the sunflower through my spiritual eyes.

This book is part math, but do not be afraid, it is relatively simple math in most cases and is fairly easy to understand. This book is part geometry and again, it is fairly simple geometry and does not get into the complexities that can be found within this field of study. It is also part esoteric history. And lastly, this book is part spiritual, in that through sacred geometry and mathematics we can see and understand the mind of creation which shows us there is an intelligence behind it, but more so, help us remove blocks preventing us from experiencing the love behind that which is creation.

Vortex based mathematics is a language or set of rules established within numbers, math and geometry itself, and is inherent within all aspects of creation, from the fabric of all matter, energy, space and time, to the creation of atoms, molecules, cells, planets, plants, stars, DNA, and living organisms. Vortex based mathematics is the coding language behind all fundamental geometries within the universe and nature. Whether it is in the form of electromagnetism, gravity, light or sound, it all follows the rules established through vortex based mathematics and is an extension of, or prerequisite to all sacred geometry, which is the study of numbers, math and geometry found inherent within nature and the universe.

Behind every sunflower there is a story to be told. It is not just a story of numbers, math and geometry, it is a story of beauty, perfection, symmetry and logic. Behind every sunflower are footprints that when followed lead to a whole new level of understanding of our physical universe and ourselves.

This book contains the hidden sacred geometry and mathematics that can be found when applying vortex based mathematics to fundamental mathematical and geometric principles found in nature and the universe. This book shares the discoveries I have made while learning about it, which contributes to the advancement of vortex based mathematics. It is also evidence in validating the work of Marko Rodin, the discoverer of vortex based mathematics.

Like the blueprints for a holographic image, we see the image but not the blueprints that make up that image. We see the image but either do not understand, or we ignore the coding language behind it. Marko Rodin discovered the evidence of this coding language, and I, along with many others, have found even further evidence validating the veracity and significance of vortex based mathematics, leading some even closer to a unified field model sought by all in physics.

Behind every sunflower are the blueprints behind creation. The numbers and their patterns and sequences can be seen and understood in every other aspect in creation because they are holographic in their nature.

Some people look behind a sunflower and see beauty while others see ugliness. Some see food and others see the potential for money or profit. I myself see the life and mind of my creator, and within it all its workings, which resonates also in you and I. I see you. I see myself. I see love…

…behind every sunflower.

Chapter 1

What is Vortex Based Mathematics?

The Fingerprint of Creation

The invisible force that is behind all of creation can be modeled in its function. When we look at numbers, they have a value, but we are not used to them having a function as well behind the order and creation of the universe. Some believe that everything evolved through random events which led to what we see and experience today. But vortex based mathematics helps us see just how perfectly it is all operating, even though it appears random and chaotic. It makes us wonder if maybe it is not all random. Maybe this shows grand design.

This force drives the atom, the stars, is behind electromagnetism, and creates all things in the universe. This invisible force really IS the universe because nothing solid really exists anyway. It is all energy, and, I believe, the so-called "solid" particle which exists in a universe of 99.999999% empty space does not even physically exist and is only a sophisticated waveform emanating from the point of all energetic source. Call it God, creation, Chi, Prana, Mana, or orgone. It is the life that gives all life.

The funny thing is, through vortex based-mathematics, we can see its fingerprint. Because numbers give rise to function, they demonstrate themselves through patterns and sequences. They demonstrate themselves through organisms, plants, flowers, molecules and atoms. And they demonstrate themselves at a larger scale through planets, solar systems, galaxies, and clusters of galaxies.

And where there are numbers, there is also geometry. It too follows this logic of perfection. In fact, when combined almost makes it unbelievable, something only God could think of. We just discovered that it was there, always was there and will always be there. It is the story that is seen in numbers and geometry, and the story they tell can explain why crystals grow the way they do; why an apple and a pumpkin grow the way they grow and why they share a shape similar to a cell; why the curve and mathematics of a spiral galaxy can also be seen in a sunflower, a pinecone, or a hurricane; why the rotations and orbits of planets synchronize into whole numbers and which are also related to harmonic geometric proportions; and why a tree branches out in the same pattern as its roots.

That shape is a toroid. Look at a donut. It has a hole at the top and at the bottom, depending on how we look at it. It is the perfect model for flow and balance. In and out, inhaling and exhaling, contracting and expanding, ingesting and excreting, pressurizing and depressurizing, positive and negative and all the while, rotating in perfect equilibrium. Like a drain in a tub, a tornado in the sky, and a black hole at the center of the galaxy, the spiraling rotation follows the path of the toroid, which I mentioned earlier is the same shape we see as a pumpkin or an apple.

These are vortexes of equal opposites, a dance of infinity. Vortex based mathematics is essentially part of vortex physics. But what differs is that vortex based mathematics demonstrates the hidden patterns and sequences of numbers that operate in tandem and explains much of vortex physics, because physics itself is vortex physics. The world just does not know it yet. But there are many scientists that now understand the importance of the toroidal dynamics. Vortex based mathematics fills in some of those empty gaps.

Marko Rodin

When Marko Rodin first discovered vortex based mathematics over forty years ago, I can only imagine the smile across his face or how he reacted when he came face to face with a view of creation that nobody had ever seen before, or at least had not seen for a very long time. I know I would probably be laughing with joy and excitement, because that is what I did when I started seeing Rodin's vortex based mathematics in discoveries of my own. It felt as if the universe, or God, was staring back at us, trying to tell us something. It was for me, confirmation beyond affirmation, like some kind of knowing. Such was this feeling, it almost seemed as if this information was coming to me in some kind of download. It came so fast and sometimes faster than I could keep up with. It was inspired by seeing others take vortex based mathematics and applying it to ratios like Phi and to the toroid shape. Once the dots were connected the rest came nearly effortlessly when testing it on drafting paper. In fact, nearly all of the designs in this book are my first attempt at drawings after a few calculations in my notebook.

I wonder if Rodin went through a similar experience. He must have, considering he discovered vortex based mathematics while studying the scriptures of the Baha'i faith. It is like the movie, "Pi," but instead of a thriller where the character loses his mind, he has an awakening instead. It is not a story of darkness, but of light.

The Doubling Circuit

So let us begin with some basic rules in the way vortex based mathematics functions and characteristics of how these numbers start to present themselves. First, there are only nine single digit numbers which are primarily used. They are 1 through 9. Anything above 9 would then be a multiple digit number and would then be calculated down to its single digital root number. For example, 10 would be adding $1 + 0 = 1$. 11 would be $1 + 1 = 2$. The number 2 would be the digital root of 11. Easy as pie.

Now let us have a little fun with numbers. But let me ask you first if you remember oh…way back to whenever you took a biology class in high school, something called cellular mitosis. It is the division of the mother cells to replicate its chromosomes by splitting themselves into two identical daughter cells to grow, and is directed by the cell's DNA.

Every one of us starts as one cell then divides into two, then four, then eight, and so on, until we are the conscious being here on this planet. This happens with all living creatures. You can actually see it for yourself. National Geographic has a video on YouTube showing the creation of a salamander from the fertilization stage to it hatching in about a six minute time-lapse video. It is absolutely amazing. The title is, *"See a Salamander Grow From a Single Cell in this Incredible Time-lapse Short Film."* [1]

Now that you have that image as a reference, let us continue with our math lesson. This is what vortex based mathematics calls the doubling circuit.

$$1 \times 2 = 2$$

$$2 \times 2 = 4$$

$$4 \times 2 = 8$$

$$8 \times 2 = 16 = 1 + 6 = 7$$

$$16 \times 2 = 32 = 3 + 2 = 5$$

$$32 \times 2 = 64 = 6 + 4 = 10 = 1 + 0 = 1$$

$$64 \times 2 = 128 = 1 + 2 + 8 = 11 = 1 + 1 = 2$$

$$128 \times 2 = 256 = 2 + 5 + 6 = 13 = 1 + 3 = 4$$

$$256 \times 2 = 512 = 5 + 1 + 2 = 8$$

$$512 \times 2 = 1{,}024 = 1 + 0 + 2 + 4 = 7$$

I bet you $5.00 that the next number in the sequence has a digital root of 5.

$$1{,}024 \times 2 = 2{,}048 = 2 + 0 + 4 + 8 = 14 = 1 + 4 = 5$$

We can now see the repeating sequence into infinity of the numbers 1, 2, 4, 8, 7 and 5. No matter how far down the table we multiply, this 6-digit sequence holds true. It even works when going in the opposite direction. Try taking 1 and halving it into infinity.

$$1 \div 2 = 0.5$$
$$0.5 \div 2 = .25 = 2 + 5 = 7$$
$$.25 \div 2 = .125 = 1 + 2 + 5 = 8$$
$$.125 \div 2 = .0625 = 0 + 6 + 2 + 5 = 13 = 1 + 3 = 4$$
$$.0625 \div 2 = .03125 = 0 + 3 + 1 + 2 + 5 = 11 = 1 + 1 = 2$$
$$.03125 \div 2 = .015625 = 0 + 1 + 5 + 6 + 2 + 5 = 19 = 1 + 9 = 10 = 1 + 0 = 1$$
$$.015625 \div 2 = .0078125 = 7 + 8 + 1 + 2 + 5 = 23 = 2 + 3 = 5$$
$$.0078125 \div 2 = .00390625 = 3 + 9 + 6 + 2 + 5 = 25 = 2 + 5 = 7$$

We now get 5, 7, 8, 4, 2 and 1, it's opposite.

Did you notice that 3, 6 and 9 were not in the sequence? It is because they serve a different function. You will soon understand. Let us take our doubling sequence and apply it to the number 3 and see what happens.

$$3 \times 2 = 6$$
$$6 \times 2 = 12 = 1 + 2 = 3$$
$$12 \times 2 = 24 = 2 + 4 = 6$$
$$24 \times 2 = 48 = 4 + 8 = 12 = 1 + 2 = 3$$
$$48 \times 2 = 96 = 9 + 6 = 15 = 1 + 5 = 6$$

$$96 \times 2 = 192 = 1 + 9 + 2 = 12 = 1 + 2 = 3$$
$$192 \times 2 = 384 = 3 + 8 + 4 = 15 = 1 + 5 = 6$$

However long we do this, the sequence will infinitely be 6, 3, 6, 3, only those two numbers. The numbers 3 and 6 are opposites but the same.

$$6 \times 2 = 12 = 1 + 2 = 3$$
$$12 \times 2 = 24 = 2 + 4 = 6$$

The only single digit number left is the number 9.

$$9 \times 2 = 18 = 1 + 8 = 9$$
$$18 \times 2 = 36 = 3 + 6 = 9$$
$$36 \times 2 = 72 = 7 + 2 = 9$$
$$72 \times 2 = 144 = 1 + 4 + 4 = 9$$

We can see that doubling 9 will always result in a digital root of itself. In fact, any number we multiply by 9 will always have a digital root of 9. I'll just throw out a random number.

$$9 \times 327 = 2{,}943 = 2 + 9 + 4 + 3 = 18 = 1 + 8 = 9$$

The Rodin Symbol

The number 9 is unto itself, whereas 3 and 6 are opposites of each other and 1, 2, 4, 8, 7 and 5 are in an infinite cycle around each other. That is what we have so far. Pretty amazing as it stands, but there is more, much more. When Rodin discovered this relationship with the doubling circuit, he discovered that it was not just fun "tricks" you can play with math. There was geometry included. The following symbol is known as "The Rodin Symbol" and contains precise geometry.

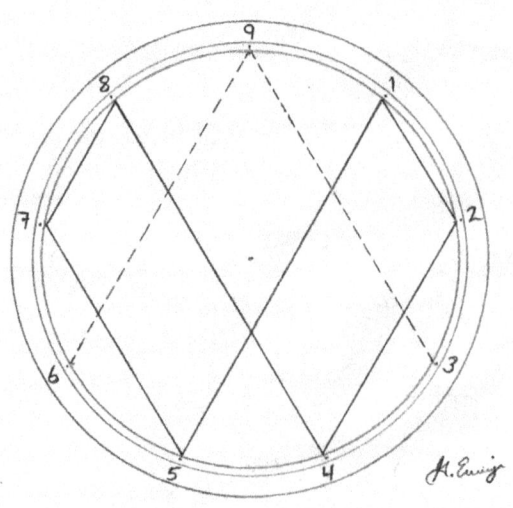

Image 1: The Rodin Symbol

Here we see the nine single digit numbers in a clockwise direction around a circle with the 9 at the top. The 9 is the central axis, the point of unity between opposites, the balance of polarity and the sum of all numbers. The 9 is the master of all other numbers and holds the balance to all form and energy. It creates all form and is all energy.

The number 9 divides into 2 numbers, becoming the duality and reciprocal balance of 3 and 6, which together equal 9. The numbers 9, 6 and 3 combined become the invisible lines of force that operate in a "higher" dimension beyond our physical dimension. If 9 is the point of source then 3 and 6 are like the bridge that completes the gap between source and the physical manifestation of 1, 2, 4, 8, 7 and 5. Thus begins the energetic dance of flow, movement and balance, contained in infinity like the path of the number 8.

This last statement sounds a little ridiculous on its face, but by the end of this book, you might second guess that assumption. That being said, let us move on.

There is no line drawn between the two numbers 3 and 6 because 9 is always the pathway between 3 and 6 as they oscillate between the two. The numbers 3 and 6 are always separated by 9. You will see evidence of this throughout the book.

The Polar Pairs

Since 9 is the axis, the balance between the two sides of the circle, we should see some evidence of this. Our first number is 1, and its equilateral opposite of 9 is the number 8. Then 2 has 7 as its opposite, 3 has 6 of course, and 4 has 5. Each of these opposites all equal 9. Let us call them polar pairs.

Now we will take each polar pair and multiply them with each number 1 through 9. We will find that their sums create mirrored numerical sequences.

The Polar Pairs of 6, 9 and 3

9

$1 \times 9 = 9$
$2 \times 9 = 18 = 1+8 = 9$
$3 \times 9 = 27 = 2+7 = 9$
$4 \times 9 = 36 = 3+6 = 9$
$5 \times 9 = 45 = 4+5 = 9$
$6 \times 9 = 54 = 5+4 = 9$
$7 \times 9 = 63 = 6+3 = 9$
$8 \times 9 = 72 = 7+2 = 9$
$9 \times 9 = 81 = 8+1 = 9$

9 has no sequence because it is the unity of itself and all

6

$1 \times 6 = 6$
$2 \times 6 = 12 = 1+2 = 3$
$3 \times 6 = 18 = 1+8 = 9$
$4 \times 6 = 24 = 2+4 = 6$
$5 \times 6 = 30 = 3+0 = 3$
$6 \times 6 = 36 = 3+6 = 9$
$7 \times 6 = 42 = 4+2 = 6$
$8 \times 6 = 48 = 4+8 = 12 = 1+2 = 3$
$9 \times 6 = 54 = 5+4 = 9$

Sequence of 6:
6 3 9 6 3 9 6 3 9

3

$1 \times 3 = 3$
$2 \times 3 = 6$
$3 \times 3 = 9$
$4 \times 3 = 12 = 1+2 = 3$
$5 \times 3 = 15 = 1+5 = 6$
$6 \times 3 = 18 = 1+8 = 9$
$7 \times 3 = 21 = 2+1 = 3$
$8 \times 3 = 24 = 2+4 = 6$
$9 \times 3 = 27 = 2+7 = 9$

Sequence of 3:
3 6 9 3 6 9 3 6 9

The Polar Pairs of the Doubling Circuit

1/8

1×1=1	1×8=8
2×1=2	2×8=16=7
3×1=3	3×8=24=6
4×1=4	4×8=32=5
5×1=5	5×8=40=4
6×1=6	6×8=48=12=3
7×1=7	7×8=56=11=2
8×1=8	8×8=64=10=1
9×1=9	9×8=72=9

Sequence of 1:
1 2 3 4 5 6 7 8 9

Sequence of 8:
8 7 6 5 4 3 2 1 9

2/7

1×2=2	1×7=7
2×2=4	2×7=14=5
3×2=6	3×7=21=3
4×2=8	4×7=28=10=1
5×2=10=1	5×7=35=8
6×2=12=3	6×7=42=6
7×2=14=5	7×7=49=13=4
8×2=16=7	8×7=56=11=2
9×2=18=9	9×7=63=9

Sequence of 2:
2 4 6 8 1 3 5 7 9

Sequence of 7:
7 5 3 1 8 6 4 2 9

4/5

1×4=4	1×5=5
2×4=8	2×5=10=1
3×4=12=3	3×5=15=6
4×4=16=7	4×5=20=2
5×4=20=2	5×5=25=7
6×4=24=6	6×5=30=3
7×4=28=10=1	7×5=35=8
8×4=32=5	8×5=40=4
9×4=36=9	9×5=45=9

Sequence of 4:
4 8 3 7 2 6 1 5 9

Sequence of 5:
5 1 6 2 7 3 8 4 9

If one side is the exact mirror opposite of the other through its math, then by mathematical terms at least we can say that the number 9 has a function or "role" as the central point of balance between two opposites or polarity. And from that we can define each other number in their opposite and call them paired numbers. Hence the polar pairs are what they are because the math and the geometry tell us so.

The Rodin Symbol and the Polar Pairs of the Doubling Circuit

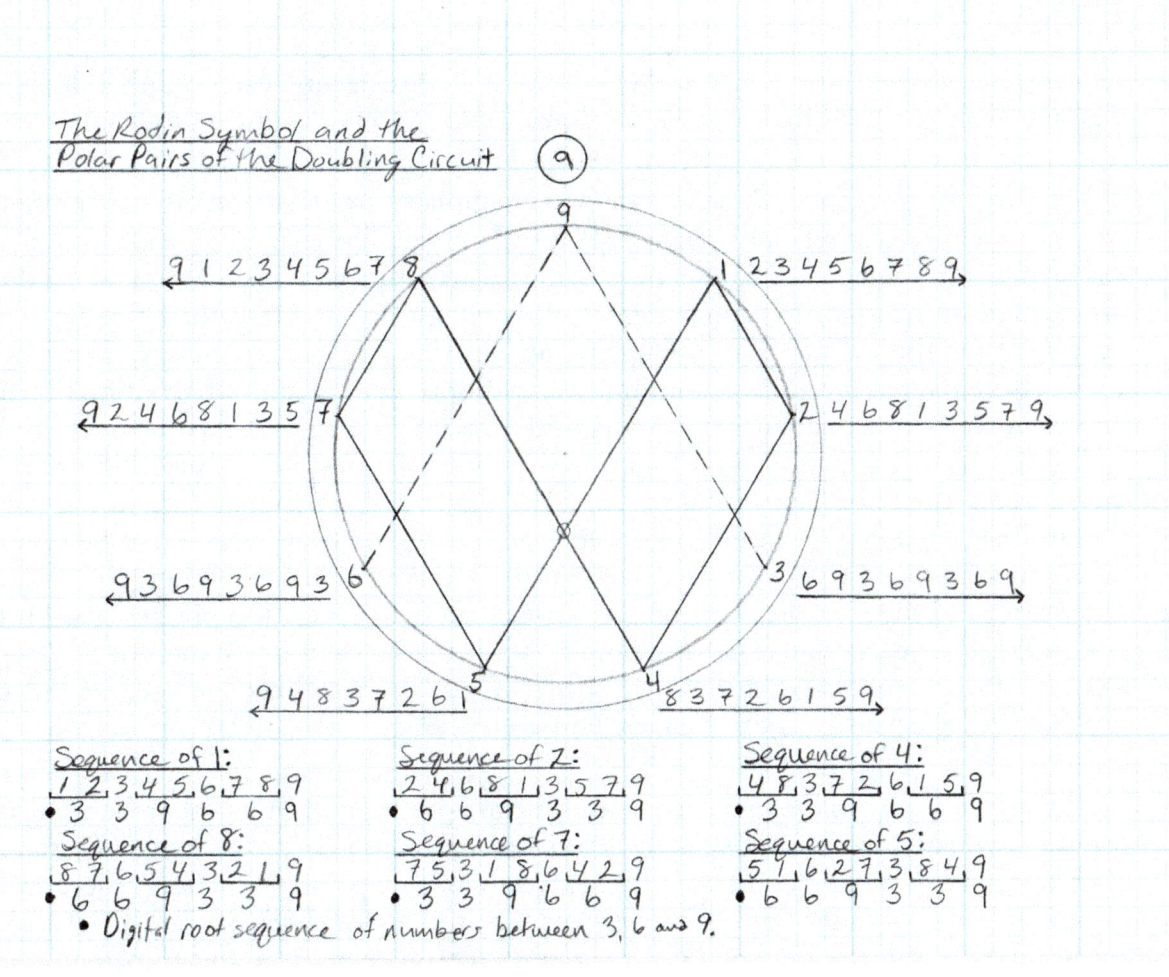

Sequence of 1:
1 2 3 4 5 6 7 8 9
• 3 3 9 6 6 9

Sequence of 2:
2 4 6 8 1 3 5 7 9
• 6 6 9 3 3 9

Sequence of 4:
4 8 3 7 2 6 1 5 9
• 3 3 9 6 6 9

Sequence of 8:
8 7 6 5 4 3 2 1 9
• 6 6 9 3 3 9

Sequence of 7:
7 5 3 1 8 6 4 2 9
• 3 3 9 6 6 9

Sequence of 5:
5 1 6 2 7 3 8 4 9
• 6 6 9 3 3 9

• Digital root sequence of numbers between 3, 6 and 9.

Roles to Play

Another way to look at the Rodin Symbol is that regarding each number, it shows why it is what it is, and the more we explore, the more obvious it becomes.

We have six sequences derived from the multiplication of each of the six numbers of the doubling circuit, and since half of them are opposites of each other, we essentially have three that have two polarities each. We also have 3, 6, and 9 which are 3 numbers apart from each other in a "family group" because their functions are separate from the 1 2 4 8 7 5 doubling circuit. The number 9 being the central point of flux between 3 and 6 creates a continuous flip-flopping of 3 and 6 becoming positive and negative, because they are opposites, and shown with a dotted line due to them not being numbers attributed in their function within the physical. If 9 is the point of unity between all numbers, the central balance, the sum of itself, and the master linchpin within it all, then it is the point of origin, pure energy or source. The number 9 then represents Spirit or God consciousness while 3 and 6, also being energetic in their extension, and together are the first steps into duality to create the physical world, are not in this physical reality. They are etheric, whereas the doubling circuit of 1, 2, 4, 8, 7 and 5 are what creates the physical world, if not the numbers themselves, but the attributes they have which can be seen in nature's design.

Imagine, if you will, energy emanating from a point, and in its movement creates a flow, or flux between opposites. It is a dance between one to the other. Written as numbers with a polar definition of positive or negative (+, -) for mere descriptive purposes helps us understand its movement. The three sequences of the six polar pairs of numbers show their energetic functions if we think of 3, 9 and 6 in that order. When we take the digital root sums of the 6-digit sequences in the previous diagram, omitting the 3, 6 and 9 (because they are functioning separately than the other six numbers), we see a pattern emerge; 1 2 3 4 5 6 7 8 9 becomes 3 3 9 6 6 9 because 1 + 2 = 3, then 3 is left alone, then 4 + 5 = 9, then 6 is left alone, then 7 + 8 = 15 and 1 + 5 = 6, and finally, 9 ends that simple sequence (see previous diagram). The same goes for every sequence, except numbers flip to their opposites; 8 7 6 5 4 3 2 1 9 its opposite sequence, becomes 6 6 9 3 3 9. And 9 has no opposites because 9 isn't split itself, just appears so to create the physical. It is unity here, that is why the path from 3 to 6 has to go through 9, or unity, to reach the other. The number 9 keeps them in motion like the tether ball player keeps the ball going around the pole while taking on the polarities of 3 and 6.

Going back to the Rodin symbol, we can see the path in the following sequence with their polarities: 3+ 3- 9- 6- 6+ 9+. Every time 3 is positive, it flips to negative, returns to 9, then to 6 being negative, which flips to positive, returning again to 9 to repeat the cycle. The sequence would also work in reverse as seen in the previous diagram.

So far, this gives us the following sequences and their opposites.

1 2 4 8 7 5 - 5 7 8 4 2 1

3 3 9 6 6 9 - 6 6 9 3 3 9, 369, 639

1 2 3 4 5 6 7 8 9 - 8 7 6 5 4 3 2 1 9

2 4 6 8 1 3 5 7 9 - 7 5 3 1 8 6 4 2 9

4 8 3 7 2 6 1 5 9 - 5 1 6 2 7 3 8 4 9

The Rodin 147 Shears

The funny thing about numbers is that they do funny things. Seriously, after a while we just laugh because we find its numerical balance, like a coherent painting showing us what a sunset looks like, we can't help but see the image. Marko Rodin derived a set of number maps comprised of 18 numbers by 18 numbers which incorporates all of these sequences, plus a few more, which I will explain. But it seems everything comes in 3's, because he found three of them in total based on combinations, each having their opposites. He calls them the 147 Shears.

These numbers and sequences fit together like a lock and key as seen in the following diagrams. The easiest way to see it is to start at the top left corner and go down 18 digits. This column will repeat the doubling circuit of 1 2 4 8 7 5 either going forward or in reverse. The second column mirrors the opposite sequence of the first. The third column repeats the 3 3 9 6 6 9 circuit, and everything after repeats to the right using the same method five more times while shifting its starting position. What assigns its starting point of the next doubling circuit going down is determined by the sequences that run 45 degrees. Interestingly, they are the multiplication sequences of the polar pairs of the doubling sequence. Also, within each set of nine digits is a square with 9 in the center. This small panel is also holographically intertwined as well as its opposite. I will show you where the three 18-digit sequences came from but a few sequences at a time, right? Here are Rodin's 147 Shears (A and B represents opposing pairs).

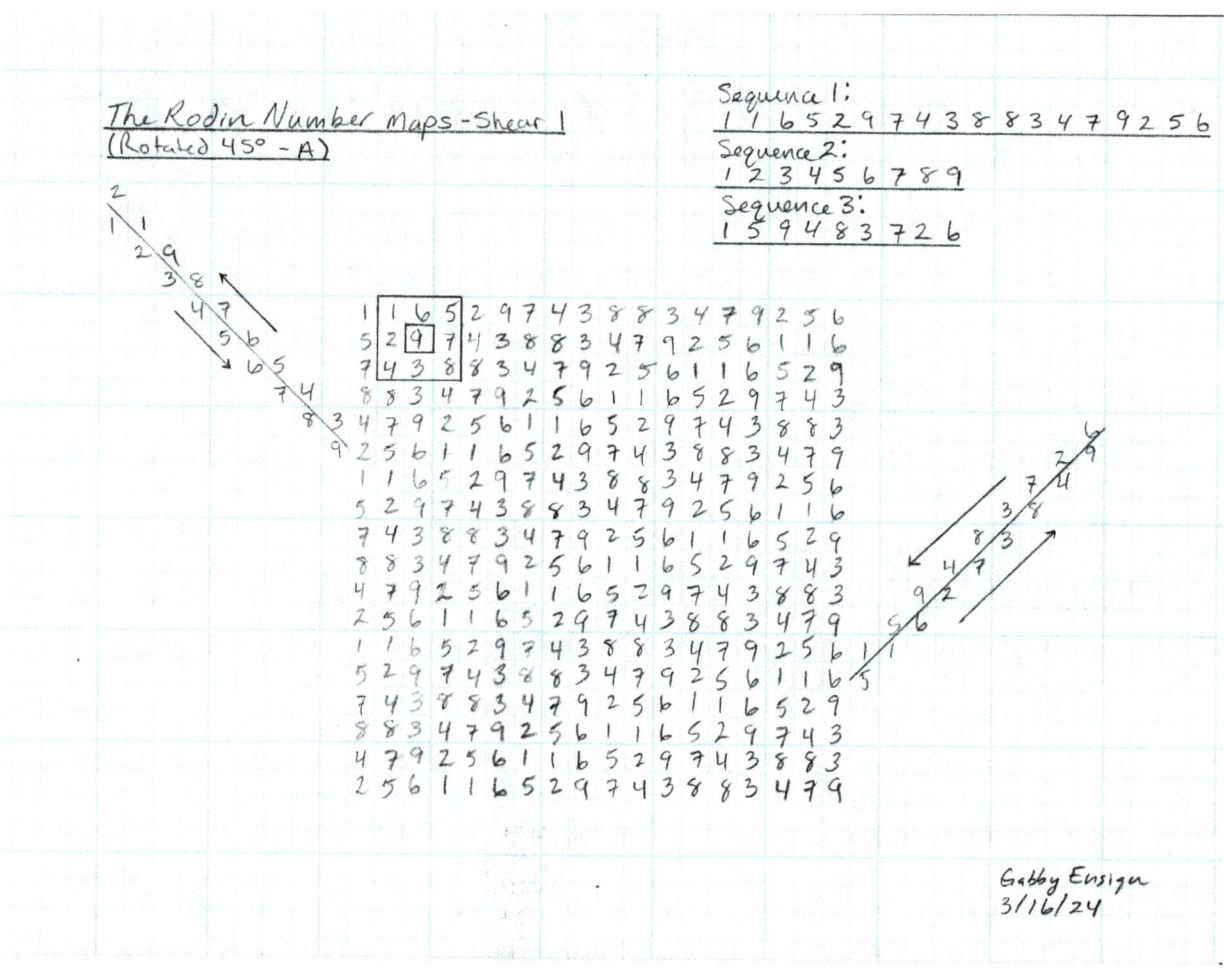

The Rodin Number Maps – Shear 1
(Rotated 45° – B)

Sequence 1:
1 1 6 5 2 9 7 4 3 8 8 3 4 7 9 2 5 6

Sequence 2:
1 5 9 4 8 3 7 2 6

Sequence 3:
1 2 3 4 5 6 7 8 9

```
        1 1 6 5 2 9 7 4 3 8 8 3 4 7 9 2 5 6
        2 5 6 1 1 6 5 2 9 7 4 3 8 8 3 4 7 9
        4 7 9 2 5 6 1 1 6 5 2 9 7 4 3 8 8 3
        8 8 3 4 7 9 2 5 6 1 1 6 5 2 9 7 4 3
        7 4 3 8 8 3 4 7 9 2 5 6 1 1 6 5 2 9
        5 2 9 7 4 3 8 8 3 4 7 9 2 5 6 1 1 6
        1 1 6 5 2 9 7 4 3 8 8 3 4 7 9 2 5 6
        2 5 6 1 1 6 5 2 9 7 4 3 8 8 3 4 7 9
        3 4 7 9 2 5 6 1 1 6 5 2 9 7 4 3 8 3
        8 8 3 4 7 9 2 5 6 1 1 6 5 2 9 7 4 3
        7 4 3 8 8 3 4 7 9 2 5 6 1 1 6 5 2 9
        5 2 9 7 4 3 8 8 3 4 7 9 2 5 6 1 1 6
        1 1 6 5 2 9 7 4 3 8 8 3 4 7 9 2 5 6
        2 5 6 1 1 6 5 2 9 7 4 3 8 8 3 4 7 9
        4 7 9 2 5 6 1 1 6 5 2 9 7 4 3 8 8 3
        8 8 3 4 7 9 2 5 6 1 1 6 5 2 9 7 4 3
        7 4 3 8 8 3 4 7 9 2 5 6 1 1 6 5 2 9
        5 2 9 7 4 3 8 8 3 4 7 9 2 5 6 1 1 6
```

Gabby Ensign
3/16/24

The Rodin Number Maps - Shear 4
(Rotated 45° - A)

Sequence 1:
1 4 9 5 8 6 7 7 6 8 5 9 4 1 3 2 2 3
Sequence 2:
1 2 3 4 5 6 7 8 9
Sequence 3:
1 8 6 4 2 9 7 5 3

```
1 4 9 5 8 6 7 7 6 8 5 9 4 1 3 2 2 3
2 2 3 1 4 9 5 8 6 7 7 6 8 5 9 4 1 3
4 1 3 2 2 3 1 4 9 5 8 6 7 7 6 8 5 9
8 5 9 4 1 3 2 2 3 1 4 9 5 8 6 7 7 6
7 7 6 8 5 9 4 1 3 2 2 9 1 4 9 5 8 6
5 8 6 7 7 6 8 5 9 4 1 3 2 2 3 1 4 9
1 4 9 5 8 6 7 7 6 8 9 3 4 1 3 2 2 3
2 2 3 1 4 9 5 8 6 7 7 9 8 5 9 4 1 3 2 2
9 4 1 3 2 2 3 1 4 9 5 8 6 7 7 6 8 5 9 4
6 8 5 9 4 1 3 2 2 3 1 4 9 5 8 6 7 7 6
7 7 6 8 5 9 4 1 3 2 2 3 1 4 9 5 8 6
5 8 6 7 7 6 8 5 9 4 1 3 2 2 3 1 4 9
1 4 9 5 8 6 7 7 6 8 5 9 4 1 3 2 2 3
2 2 3 1 4 9 5 8 6 7 7 6 8 5 9 4 1 3
4 1 3 2 2 3 1 4 9 5 8 6 7 7 6 8 5 9
8 5 9 4 1 3 2 2 3 1 4 9 5 8 6 7 7 6
7 7 6 8 5 9 4 1 3 2 2 3 1 4 9 5 8 6
5 8 6 7 7 6 8 5 9 4 1 3 2 2 3 1 4 9
```

The Rodin Number Maps - Shear 4
(Rotated 45° - B)

Sequence 1:
1 4 9 5 8 6 7 7 6 8 5 9 4 1 3 2 2 3

Sequence 2:
1 8 6 4 2 9 7 5 3

Sequence 3:
1 2 3 4 5 6 7 8 9

```
1 4 9 5 8 6 7 7 6 8 5 9 4 1 3 2 2 3
  5 8 6 7 7 6 8 5 9 4 1 3 2 2 3 1 4 9
    7 7 6 8 5 9 4 1 3 2 2 3 1 4 9 5 8 6
      8 5 9 4 1 3 2 2 3 1 4 9 5 8 6 7 7 6
        4 1 3 2 2 3 1 4 9 5 8 6 7 7 6 8 5 9
          2 2 3 1 4 9 5 8 6 7 7 6 8 5 9 4 1 3
            1 4 9 5 8 6 7 7 6 8 5 9 4 1 3 2 2 3
              5 8 6 7 7 6 8 5 9 4 1 3 2 2 3 1 4 9
                7 7 6 8 5 9 4 1 3 2 2 3 1 4 9 5 8 6
                  8 5 9 4 1 3 2 2 3 1 4 9 5 8 6 7 7 6
                    4 1 3 2 2 3 1 4 9 5 8 6 7 7 6 8 5 9
                      2 2 3 1 4 9 5 8 6 7 7 6 8 5 9 4 1 3
                        1 4 9 5 8 6 7 7 6 8 5 9 4 1 3 2 2 3
                          5 8 6 7 7 6 8 5 9 4 1 3 2 2 3 1 4 9
                            7 7 6 8 5 9 4 1 3 2 2 3 1 4 9 5 8 6
                              8 5 9 4 1 3 2 2 3 1 4 9 5 8 6 7 7 6
                                4 1 3 2 2 3 1 4 9 5 8 6 7 7 6 8 5 9
                                  2 2 3 1 4 9 5 8 6 7 7 6 8 5 9 4 1 3
```

The Rodin Number Maps – Shear 7
(Rotated 45° – A)

Sequence 1:
1 7 3 5 5 3 7 1 9 8 2 6 4 4 6 2 8 9

Sequence 2:
1 8 6 4 2 9 7 5 3

Sequence 3:
1 5 9 4 8 3 7 2 6

```
          1 7 3 5 5 3 7 1 9 8 2 6 4 4 6 2 8 9
          2 8 9 1 7 3 5 5 3 7 1 9 8 2 6 4 4 6
          4 4 6 2 8 9 1 7 3 5 5 3 7 1 9 8 2 6
          8 2 6 4 4 6 2 8 9 1 7 3 5 5 3 7 1 9
          7 1 9 8 2 6 4 4 6 2 8 9 1 7 3 5 5 3
          5 5 3 7 1 9 8 2 6 4 4 6 2 8 9 1 7 3
          9 1 7 3 5 5 3 7 1 9 8 2 6 4 4 6 2 8 9
          6 2 8 9 1 7 3 5 5 3 7 1 9 8 2 6 4 4 6
          4 4 6 2 8 9 1 7 3 5 5 3 7 1 9 8 2 6 4 4
          8 2 6 4 4 6 2 8 9 1 7 3 5 5 3 7 1 9 8
          7 1 9 8 2 6 4 4 6 2 8 9 1 7 3 5 5 3
          5 5 3 7 1 9 8 2 6 4 4 6 2 8 9 1 7 3
          1 7 3 5 5 3 7 1 9 8 2 6 4 4 6 2 8 9
          2 8 9 1 7 3 5 5 3 7 1 9 8 2 6 4 4 6
          4 4 6 2 8 9 1 7 3 5 5 3 7 1 9 8 2 6
          8 2 6 4 4 6 2 8 9 1 7 3 5 5 3 7 1 9
          7 1 9 8 2 6 4 4 6 2 8 9 1 7 3 5 5 3
          5 5 3 7 1 9 8 2 6 4 4 6 2 8 9 1 7 3
```

Gabby Ensign
3/16/24

The Rodin Number Maps - Shear 7
(Rotated 45°-B)

Sequence 1:
1 7 3 5 5 3 7 1 9 8 2 6 4 4 6 2 8 9
Sequence 2:
1 5 9 4 8 3 7 2 6
Sequence 3:
1 8 6 4 2 9 7 5 3

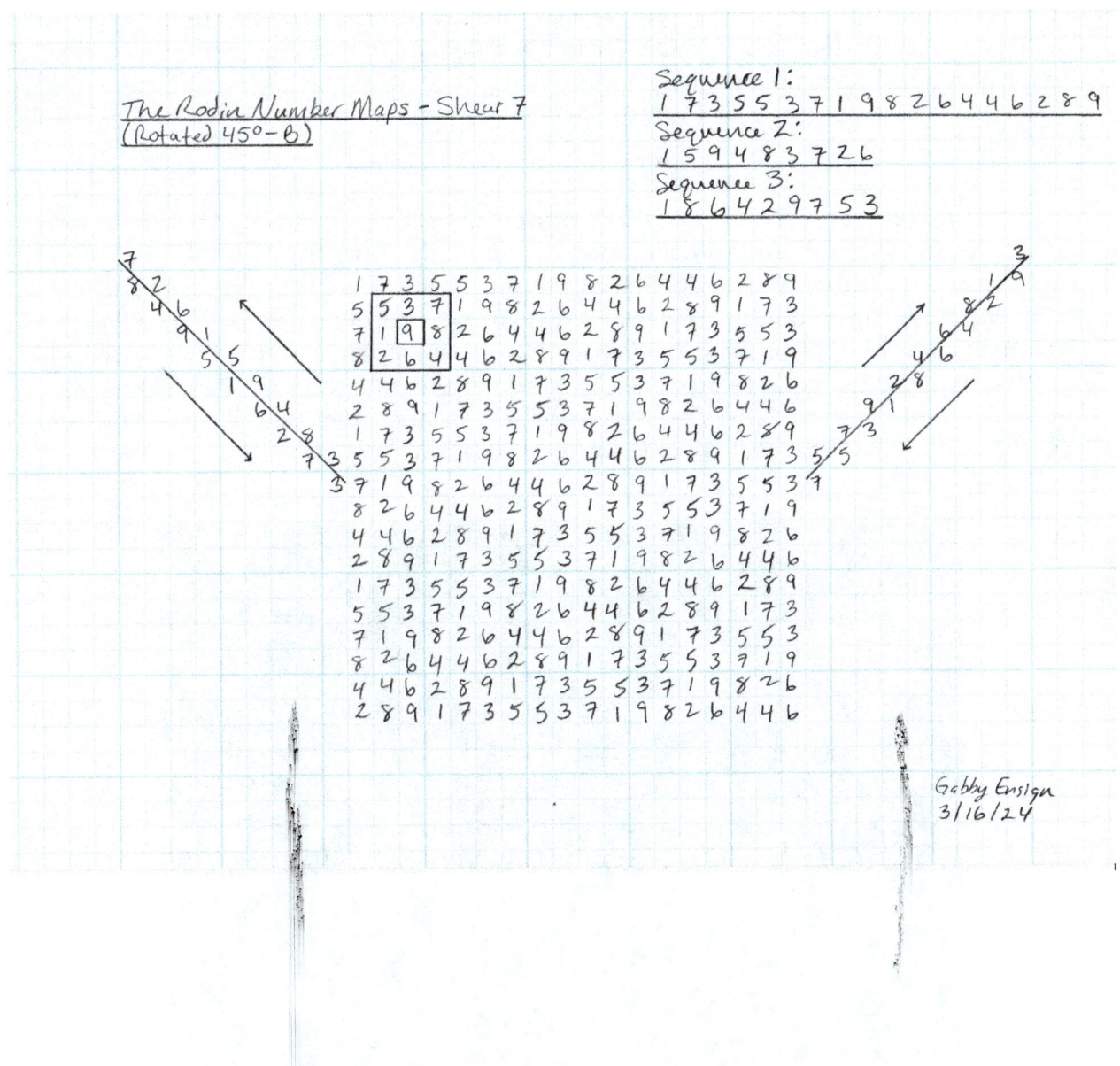

Gabby Ensign
3/16/24

The reason we only have three grid maps and their opposites using this method is because that is all the combinations there are which result in coherent numerical alignments with all the key sequences. They demonstrate such balance because numbers work individually and as a whole to show this level of intertwined connectedness.

Nikola Tesla said, "If you only knew the magnificence of the 3, 6 and 9, then you would have the key to the universe." Vortex based mathematics helps us see what Tesla knew over a century ago. We see there is a uniqueness with 3, 6 and 9 with their math, with their relationship to each other in the number maps and their position on the Rodin Symbol. This is only the beginning.

The Toroid

So, at this point we have been talking mostly about the 9 single digit numbers and their relationships with each other through basic math, using their digital root sums, and the Rodin Symbol. Where do vortexes fit into this? Would you like an apple? Cut it in half from top to bottom. This is a side view of a toroid cut in half. The energy that makes up the apple is like a field moving from one point, through the center and out the other side to go back around the outside to the point inward in a rotating and curving infinite loop of a system in perfect balance and efficiency. That is why our planet's electromagnetic field is in the shape of a toroid.

Image 2: Earth's Magnetic Field

In fact, the toroid shape is one of the most fundamental shapes in nature. The movement of a tornado follows these same vortex principles. It is the same for the movement of a drain, a hurricane, or a spiral galaxy. As mentioned before, many plants and their fruit show this shape, or the attributes of its mathematical features. The vortex movement follows a curve that conforms to the golden number Phi of 1.618...to infinity. This same curved spiral is very prominently seen in pinecones and sunflowers, but has its roots in nearly every aspect of creation down to the lengths of our bones in relationship to each other.

Image 3 – 10: Toroid Examples

In the next image you can see what Rodin and his colleagues did with the Rodin number grid maps (which I like to call them) by placing them on the surface of a toroid to make the Shears loop with themselves in an infinite sequence like the natural world already does. But instead of squares on a two-dimensional flat plane, they become diamonds that follow the directional movement of a three-dimensional form. The diamonds are of course representational of the properties of numbers behind the movement of form, and are always aligned towards the center.

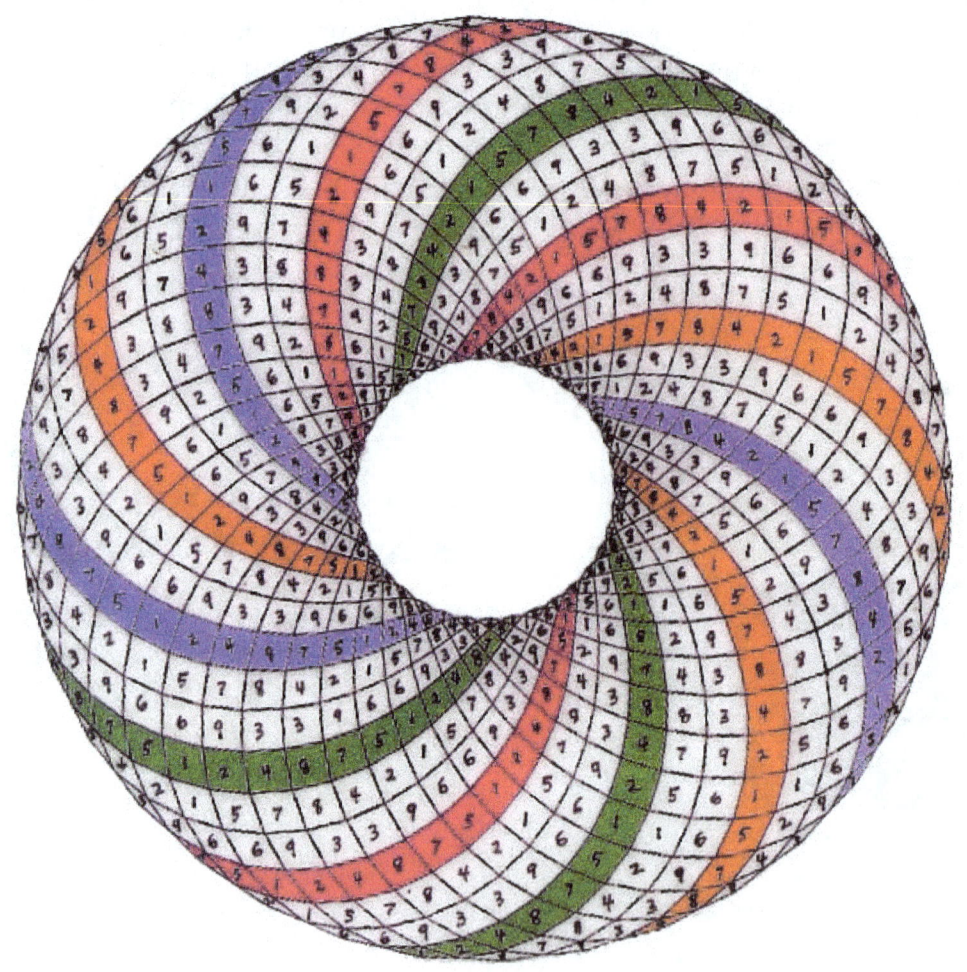

Image 11: The Rodin Toroid

Rodin, (as well as his colleagues and I), understands the importance the toroid shape has on the creation of form. There is a fast-growing awareness that this shape is the energetic system that operates at an atomic level to create particles, atoms, and molecules as well. "As above, so below" is what science is now showing us. If form works in such a balanced system in infinite and opposite sequences, the common relationship between the two couldn't be more apparent. Somehow there must be a relationship. Vortex based mathematics begins with this premise, that numbers play an important role within all functions related to creation. It is not so different than Pythagoras believing geometry held the keys to the mind of God, and to the building blocks of the universe in the 6th century B.C. Except now we have ways to observe, measure and test this theory and it turns out to be absolutely correct.

The Gift of Knowing

When talking about fundamental geometry as the actual building blocks for creation to someone who has done their research, it is no longer an argument, it is a fact to them. Numbers, like geometry also do not lie. They simply are. They are intimately connected to geometry in all of its forms, and reflect its harmony and balance. The discovery of vortex based mathematics by Marko Rodin shows us that numbers follow certain patterns in many astounding ways, and I along with others are attempting to understand its role in how the universe works. If you have a good understanding of algebra, I encourage you to look into the work of mathematician, Greg Volk. He can show the spacial relationships of this math through algebraic equations, and gives further validation of the use of vortex based mathematics.

Although volumes of information on vortex based mathematics can be found online,[2] I have given you just enough to understand what comes next. If you have no idea what sacred geometry is or have never heard the term, I suggest you start there so you can fully understand and appreciate what this book has to offer. From me to you in the simplest terms, I will try to give you the gift of knowing that the universe is a product of design and perfectly functioning systems, with such efficiency at a mass scale that you walk away with a better sense of something greater behind it all.

Chapter 2

The Divine Proportion and Vortex Based Mathematics

Phi 1.618

The first time I learned about Phi, I was about 17 years old, living with my dad. He had purchased a study course on cassette tape. The course was a type of spiritual development and abundant success program. The logo the company used was the Phi spiral diagram.

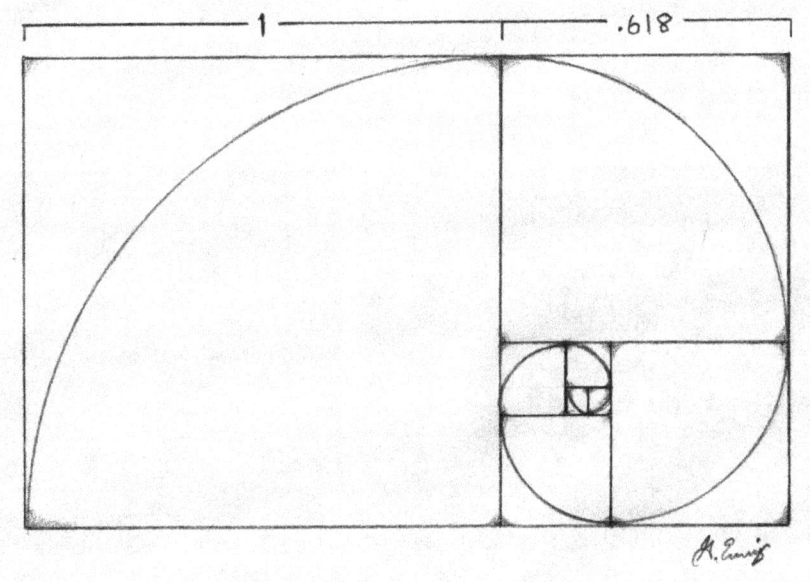

Image 12: The Phi Spiral

I learned what it was and have been fascinated with it ever since. Here is the numerical sequence.

$$0 + 1 = 1$$
$$1 + 1 = 2$$
$$2 + 1 = 3$$
$$3 + 2 = 5$$
$$5 + 3 = 8$$
$$8 + 5 = 13$$

$$13 + 8 = 21$$
$$21 + 13 = 34$$
$$34 + 21 = 55$$

...and so on.

The rule is to add the sum of the number to the previous one to get the next number in the sequence, starting with 0 + 1. Eventually, we would end up with a ratio of 1.618033989… if you were to divide any two numbers next to each other in the sequence.

The diagram is created by this ratio using squares rotating outwardly which results in a curved spiral using the arc of each square. Here are a few examples of this divine ratio expressing itself in nature.

Image 13 – 17: Phi in Nature Examples

This ratio is so pervasive that it goes by many names, a few of which I have already mentioned. The names are of course, Phi, the golden ratio, the golden mean, the golden number, The Fibonacci Sequence, the divine number, the divine proportion, the golden spiral, etc.

The ratio has been known for centuries, going back to at least 1202 A.D., when Leonardo Fibonacci, an Italian mathematician, rediscovered its importance and named it The Fibonacci Sequence,[3] although there was evidence of it being known by Indian Mathematicians over a thousand years earlier.[4]

You can find evidence of this ratio in ancient structures like the Parthenon, the Great Pyramid, Notre Dame, and The Taj Mahal, and it was used by famous artists like Leonardo Da Vinci in the Mona Lisa, The Last Supper and the Vitruvian Man.[5]

Image 18 – 22: Phi in Art and Architecture

I encourage you to learn more about this amazing number because it seems to pop up in so many places. From the vast size of the spiraling arms of our galaxy, down to the spiral helix of our DNA, it alone is enough to postulate intelligent design. That being said, let us get into some examples of Phi in vortex based mathematics.

The Rodin Phi Diagram

The following diagram was derived from the Rodin Symbol and the Phi diagram. I did not know where to start in order to find this alignment until I was flipping through the book, "The Ancient Secret of the Flower of Life; Vol. 2," by Drunvalo Melchizedek. I came across a layout of the Great Pyramids at Giza in Egypt, which depicted a Phi spiral aligning the pyramid complex and the Sphinx to 30° from the heliacal rising of Sirius. So, I decided, why not give it a try with the Rodin Symbol?

I got out my drafting tools and compass, and I made a perfect diagram of the Rodin Symbol. I then made a horizontal line at 0°/360° and another at 30°, then constructed the Phi diagram in proportion to the center and I crossed my fingers as I started drawing it out.

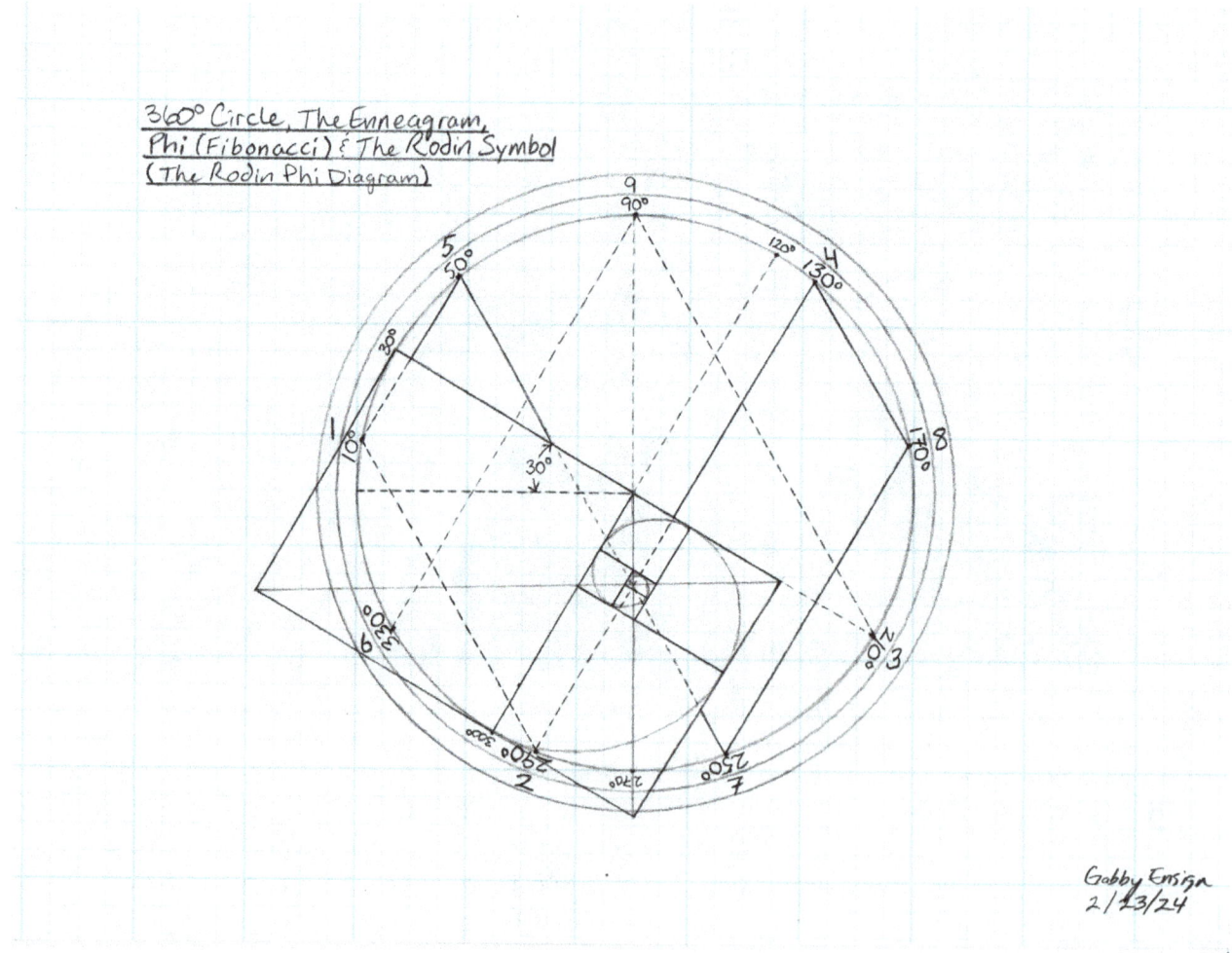

To my utter amazement, it lined up! The point in which the four intersecting lines of the W cross is what Rodin calls the emanation point of eternity, the point of unity, source energy, Spirit, Mana, and Chi; you get the idea. It is the point from which all of physical creation emanates. Well, isn't it interesting that Phi, being the ratio expressed in seemingly all aspects of creation, is emanating from Rodin's "point of creation" in perfect alignment? If you were to follow the spiral inward in a computer model, you would find it infinitely getting closer to this point but never reaching it, always spiraling closer and closer and closer. We can appropriately say in this case, "All comes from the Infinite."

I call this diagram the Rodin Phi Diagram. It was inspired by Nassim Haramein, from the Resonance Academy. On a video, Nassim is talking to his friends about the Phi spiral proportions lining up with the Rodin Symbol, and he showed a CAD drawing somewhat depicting it. I showed the Rodin Phi Diagram to Marko Rodin, and he said that originally the idea was given to Nassim by a friend and colleague of his, Andre Nugess, but he never gave Andre credit for it. If it is true or not, I can't say, nor can I say that the design is mine, nor Andre's for that matter. It is an idea, a design, a concept, which precedes even the Big Bang. But this is only a representation, if you will, of the actual process. As members of King Arthur's Court said in Monty Python's, The Holy Grail, *"Camelot! ...Camelot! ...Camelot! Shhhhh, it is only a model."*

The Rodin Phi Diagram was created much later than my other diagrams, but due to its significance, I felt it appropriate to show it in the beginning, because at this point out, you can consider the rest of this book as having a relationship in some way to Phi. They coexist together, because Phi is a function of the system. So, consider this chapter the beginning of establishing Phi as a rotating three-dimensional and even multidimensional inward and outward curve, as seen in a vortex, for example. This is how Chi moves in vortex based mathematics through its function as a toroidal field of energy.

Phi, being just a representation of the proportions we see in nature, is also a phenomenon seen in numerical diagrams in vortex based mathematics. This actually takes us to the beginning of my journey into my own discoveries.

Gabby's Rodin Flower

A couple of weeks after I learned about vortex based mathematics, when Marko Rodin wandered into my band room back yard, I had an idea regarding sequences and their opposites, or mirror. I thought if I took the Rodin symbol and applied the same sequence and diagram to its polar opposite of the number 9, would that give me any insights? I decided to test my idea on the flower of life, a pattern familiar to anyone studying sacred geometry, because it is also referred to as the creation pattern, due to its perfect relationship to all geometric solids, including ratios like Phi. A perfect test!

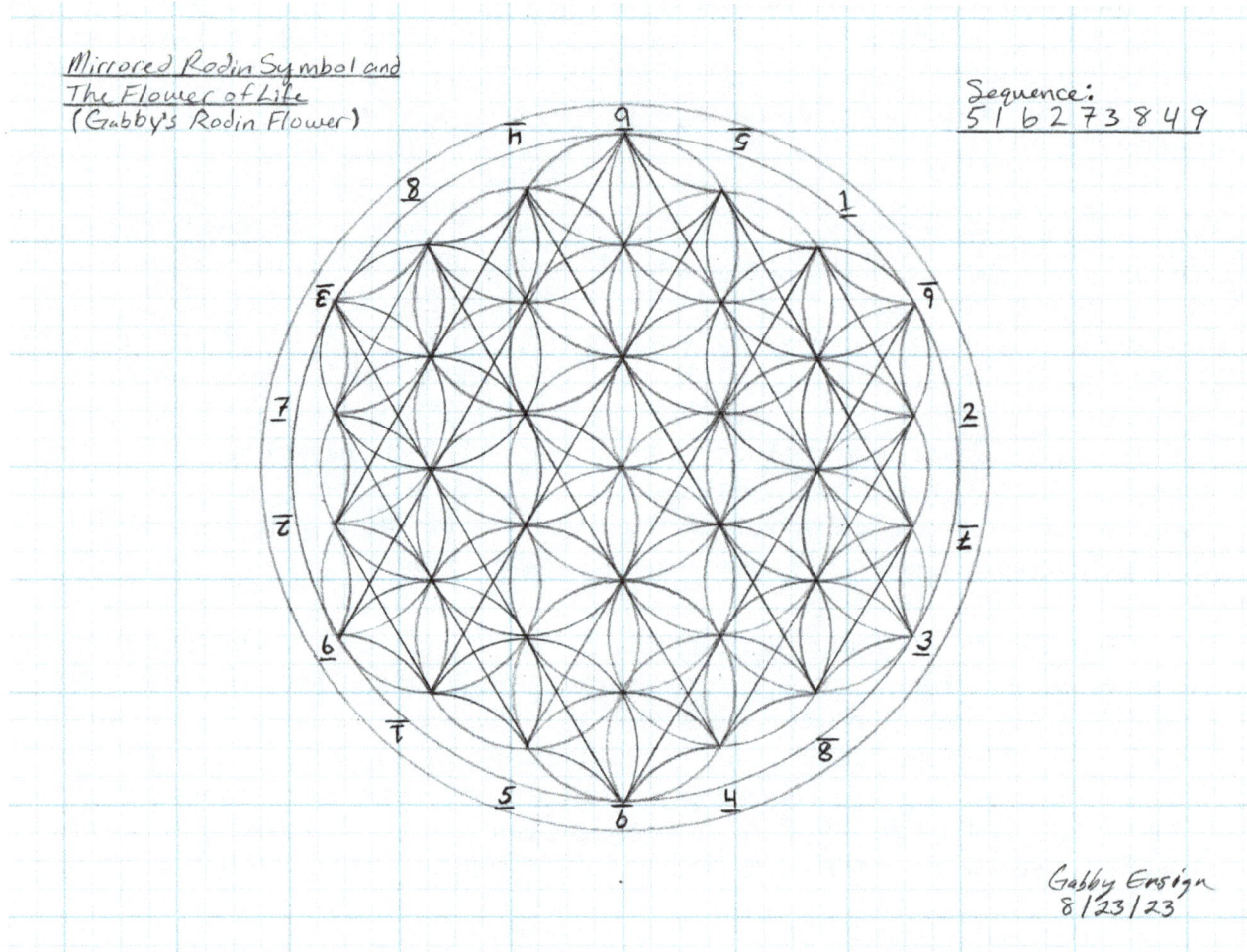

Other than some minor imperfections in my drawing, the 18 points of the Rodin Symbol and its opposite make direct lines of paths between the flower of life pattern in its second phase of development as the egg of life. (The egg of life phase comes after the seed of life phase, and before the flower of life phase.) This phase creates 18 outer points of connection and is the final stage of its completion of the first 8 cells of mitosis. It is also the third set of circles outward with the center circle being the first.

Gator's Donut

Besides the fact that my test seemed to satisfy the geometry of the Rodin Symbol and its opposite to the flower of life pattern, it also created a sequence of numbers I would soon find was directly related to another major vortex based mathematics discovery by Mike Trefrei shown on YouTube@Inphiknitfractal.[6] He calls it Gator's Donut and it originates with the sequence 5 1 6 2 7 3 8 4 9, which I later found out was the 4-5 polar pairs sequence. Every other number on my diagram is upside down because it is showing its relationship to the opposite 1 through 9 sequence moving opposite and in-between. But it is the exact same sequence around a circle. It is also a sequence within the Rodin 147 Shears.

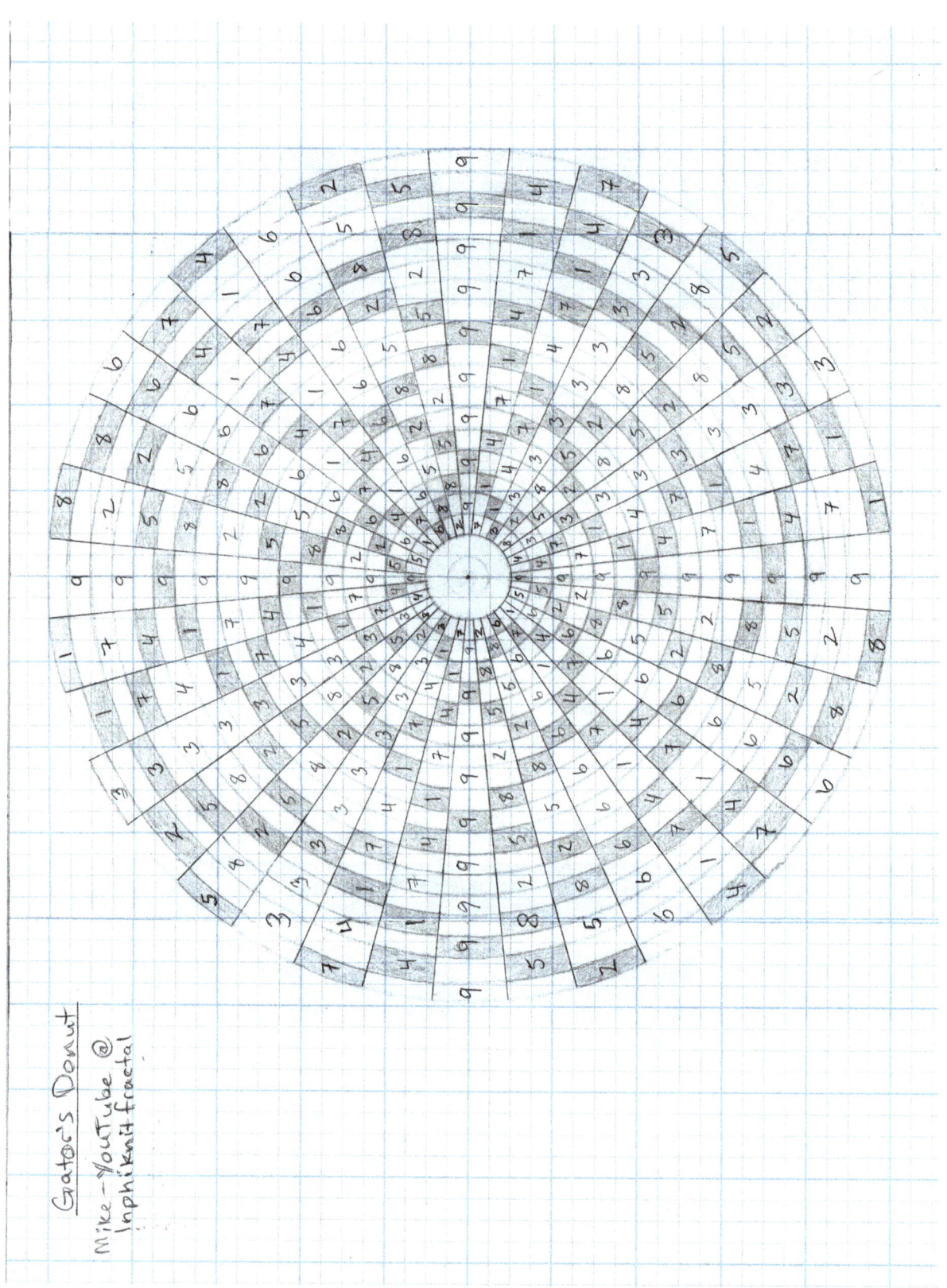

Gator's Donut

Mike - YouTube @ Inphiknit fractal

Looking at Gator's Donut, if you look to the center of the circle, you will see this sequence in its first step. The second step would be the next outer ring, which is the digital root sum of the numbers neighboring each other in the first step. For example, 9 + 5 = 14 equals 5, so 5 would be between 9 and 5 on the second outer step. Then 5 + 1 = 6, 1 + 6 equals 7, 6 + 2 = 8, 2 + 7 = 9, and so on until the second step shows a sequence of 1 2 3 4 5 6 7 8 9. Do this again to calculate the third step and we get the sequence, 2 4 6 8 1 3 5 7 9. The fourth step is the same as the first step except in its reverse opposite direction. The same goes for the fifth and sixth steps, then repeats.

These three sequences are from the Rodin symbol polar pairs, and their opposites, repeating outwardly from themselves through the digital root sums of themselves as if the three were intimately intertwined, just like the Rodin number grid maps of the Rodin 147 Shears.

Along with these three sequences repeating, outwardly, there are three other sequences and their opposites, extending outward at a diagonal direction from the center. I have shaded in a few so you can clearly see the Phi spiral curving outwardly from the center. The three sequences are exactly the same sequence of numbers going horizontally across the numerical grid maps of the Rodin 147 Shears, except the 3's, 6's and 9's have shifted by one position of seeing the 3 3 9 6 6 9 set of numbers representing its own sub-sequence interwoven into the main sequence of 18 digits. It makes sense because 3, 9 and 6 operate separately from 1, 2, 4, 8, 7 and 5. Here are the next three sequences which you similarly may have seen in Rodin's number grid maps for his 147 Shears.

Sequence 1:
1 1 3 5 2 3 7 4 9 8 8 6 4 7 6 2 5 9

Sequence 2:
1 4 6 5 8 9 7 7 3 8 5 3 4 1 9 2 2 6

Sequence 3:
1 7 9 5 5 6 7 1 6 8 2 9 4 4 3 2 8 3

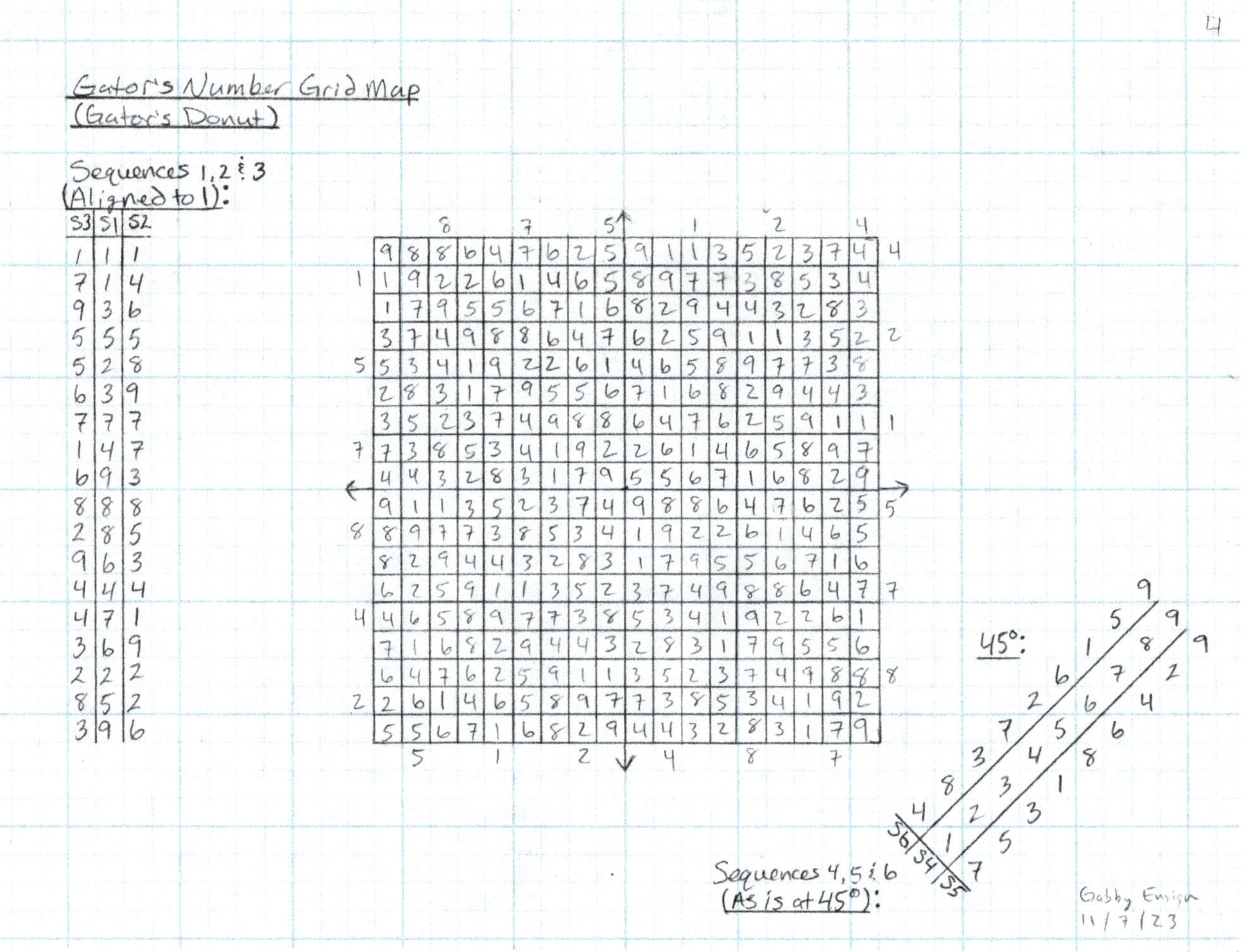

The three 18-digit sequences repeating in Rodin's grid map are as follows:

Sequence 1:
1 1 6 5 2 9 7 4 3 8 8 3 4 7 9 2 5 6
Sequence 2:
1 4 9 5 8 6 7 7 6 8 5 9 4 1 3 2 2 3
Sequence 3:
1 7 3 5 5 3 7 1 9 8 2 6 4 4 6 2 8 9

Now can you see how the 3 3 9 6 6 9 placement has moved its position when looking at every third number? Keep this in the back of your mind for now. Next, you'll notice that 45° in one direction of their numerical grid map are the polar pairs of 1 through 9 which are 1 2 3 4 5 6 7 8 9, 2 4 6 8 1 3 5 7 9 and 5 1 6 2 7 3 8 4 9. This shows symmetry and correlates perfectly. So let us see what else this toroidal grid map shows us.

Looking at the other 45° angle to its opposite, you can easily see a row of 9's. From there each row repeats as follows.

5 8 2 5 8 2...
5 8 2 5 8 2...
6 6 6 6 6 6...
7 4 1 7 4 1...
1 7 4 1 7 4...
6 6 6 6 6 6...
8 2 5 8 2 5...
2 5 8 2 5 8...
9 9 9 9 9 9...
4 1 7 4 1 7...
4 1 7 4 1 7...
3 3 3 3 3 3...
2 5 8 2 5 8...
8 2 5 8 2 5...
3 3 3 3 3 3...
1 7 4 1 7 4...
7 4 1 7 4 1...
9 9 9 9 9 9...

Again, the first number in each of these sequences is one of the three 18-digit sequences of the horizontal plane. Now we can plainly see the 3, 9 and 6 as a separate function of the numbers.

Every fourth number of the 18-digit sequences shows the 1 2 4 8 7 5 circuit or the 3 3 9 6 6 9 circuit as Rodin calls them. It also shows the 1 4 7 and 2 5 8 family number groups, which I will explain more about in the following chapters. Needless to say, Gator's Donut has vortex based mathematics written all over it. It also inspired me to continue playing around with numbers because I found something so incredible that if I wasn't already convinced of the depth and magnitude of vortex based mathematics, this would have done it for me.

The Root Sum of Phi

Shortly after discovering Gator's Donut, I came across a video by *@WhyPhi?* on YouTube explaining the Fibonacci sequence and its relationship to vortex based mathematics.[7] They took the numbers in the sequence and reduced them down to their digital root sums as far as it took them to realize that a pattern started to emerge. A 24-digit sequence would repeat itself into infinity.

1 1 2 3 5 8 4 3 7 1 8 9 8 8 7 6 4 1 5 6 2 8 1 9 1 1 2 3…

The following diagram shows the sequence around 24 points on a circle. The outer larger circle shows the digital root sums of each set of three numbers.

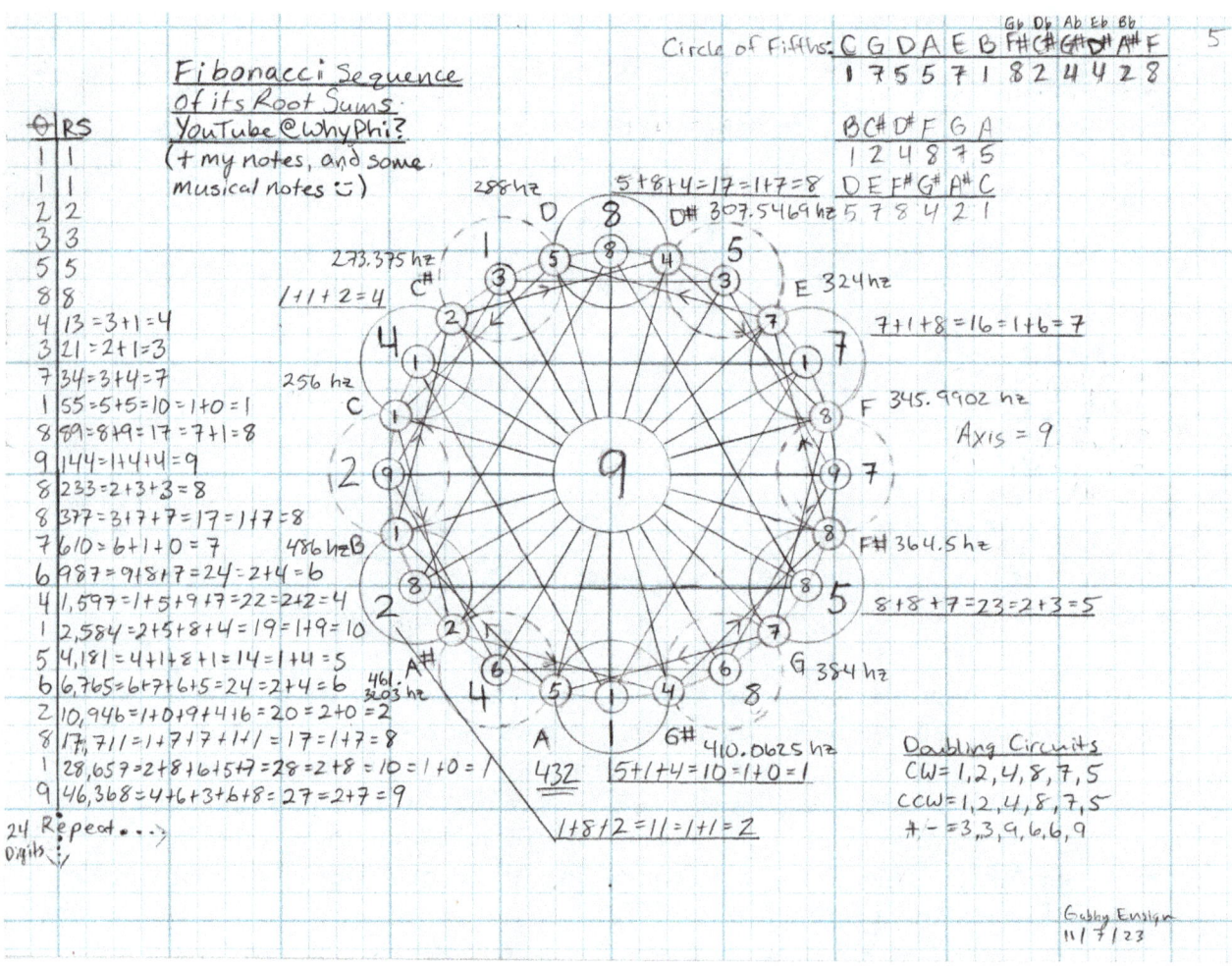

There are two doubling circuits running in opposite directions from each other. Clockwise, we see every other number of the outer larger circles repeating 1 2 4 8 7 5, and in the opposite direction of every other number, repeating 1 2 4 8 7 5. I've separated the two opposing circuits by solid and dotted lines forming the circle.

The Diamond of Fifths

On the left, you can see the math for the digital root sums that creates the sequence. Elsewhere I have assigned musical notes and their hertz relationships based on the 12 notes of the musical scale. I'll go into more detail about that but first I would like to bring your attention to the 12-digit sequence in the upper right. Does it look a bit familiar? Since I now understand that 3, 9 and 6 operate separately from the others, I treated the 24-digit Phi sequence as two separate 12-digit sequences operating apart from each other, which seem to make sense because 3, 9 and 6 only appeared in the center of the large circles and none of the 12 smaller circles in between them. So, I composed an experiment, and its results painted a clear picture of vortex based mathematics being clearly apparent within the numbers of Phi. The two sequences are 1 2 5 4 7 8 8 7 4 5 2 1 and 1 3 8 3 1 9 8 6 1 6 8 9. For now, let's look at the first sequence with no 3's, 6's or 9's, because why not, we start with what we can see, and since 3 9 6 are non-physical in their nature, I will start with the sequence containing only numbers from the doubling circuit of 1 2 4 8 7 5. You'll notice right away in the following diagram that the 12-digit sequence at the top creates a geometric pattern. It reaches one of the 12 points in perfect symmetry. There are other ways to get to every point only once without hitting another more than once until repeating the cycle, but there are only a few which follow perfect symmetry and are not random movements or incoherent pathways.

This sequence, 1 7 5 5 7 1 8 2 4 4 2 8, I discovered is exactly one of Rodin's grid map sequences, and one of Gator's Donut sequences, but without the 3 3 9 6 6 9 circuit running through it. Well, 18 digits minus this 6-digit circuit equals another 12-digit sequence. Amazing! And right off the Phi sequence itself! At this point, my jaw dropped even further.

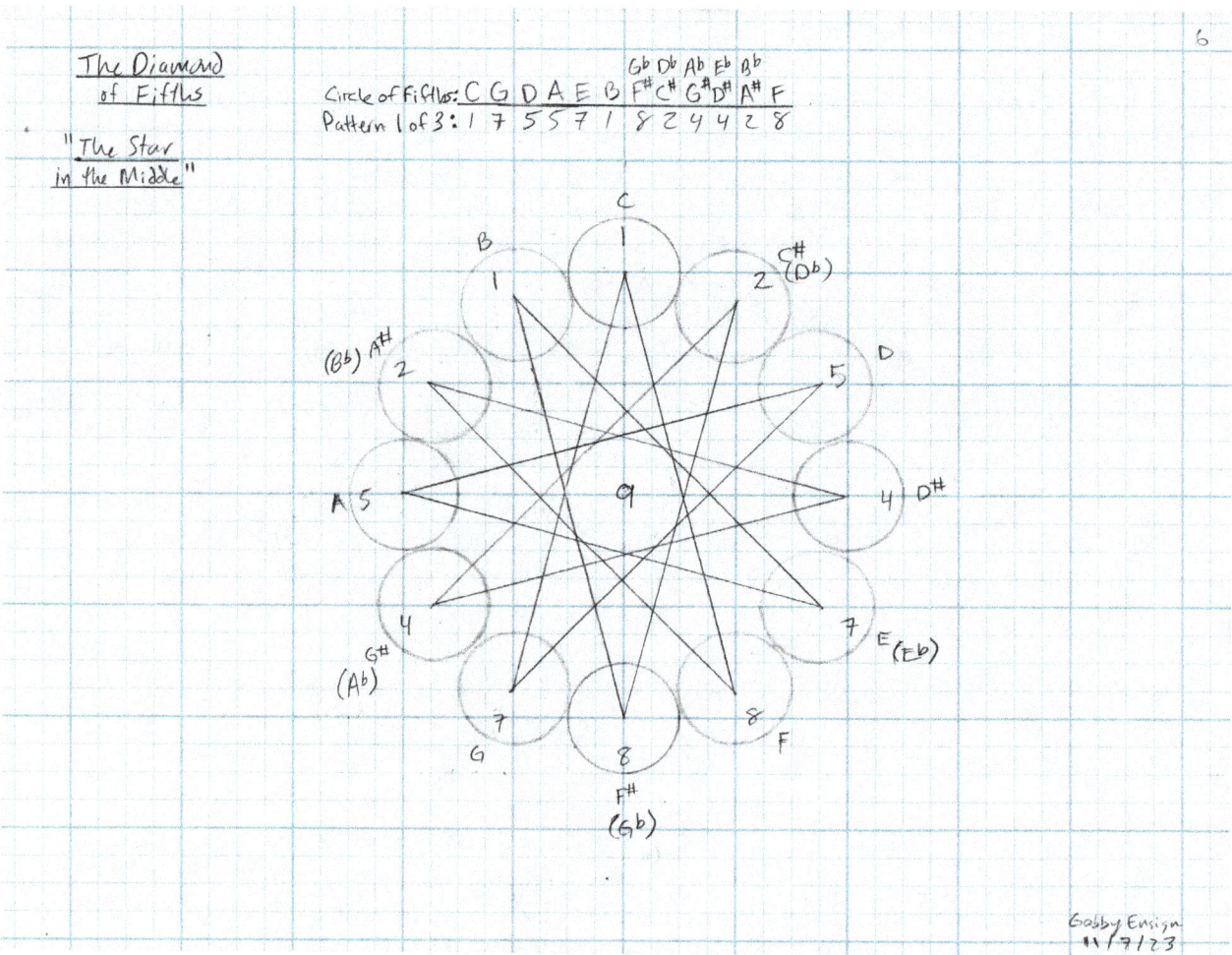

    I immediately recognized this pattern as the circle of fifths geometry in musical theory. The circle of fifths is a diagram that was made in order to travel from any notes in the chart to another and maintain a sense of sounding pleasing to the ear. For whatever reason each note in relationship to each other just sounds good together, as opposed to arrangements that deviate from this pattern. It is seen extensively, it seems, in many well-known songs. Here is the layout of the notes in the circle of fifths.

    C G D A E B G♭ D♭ A♭ E♭ B♭ F

    If we look at this sequence around the circle, you will notice the two counter-rotating doubling circuits of 1 2 8 4 7 5 again like we saw with the sum of 3 numbers from the original 24-digit sequence. It seems they are holographic in their nature like a set of Russian dolls. This is yet another testimony to the significance of Phi in vortex based mathematics as a real observation or expression through numbers, because Phi is basically in everything.

Since I had a symmetrical pattern from the first sequence derived from the Rodin 147 Shears, I wondered if the other two would show any symmetrical patterns as well. Sure enough, they did.

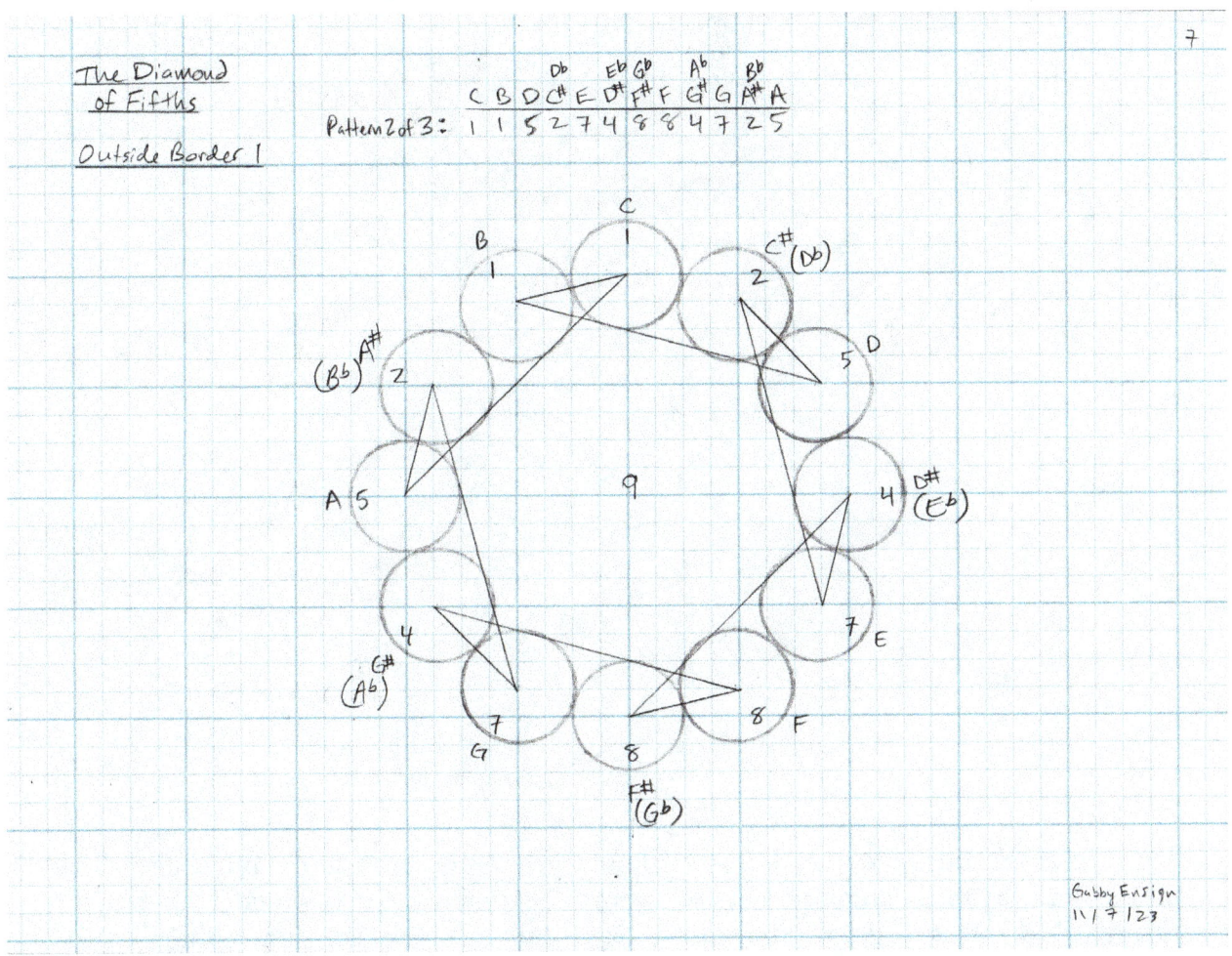

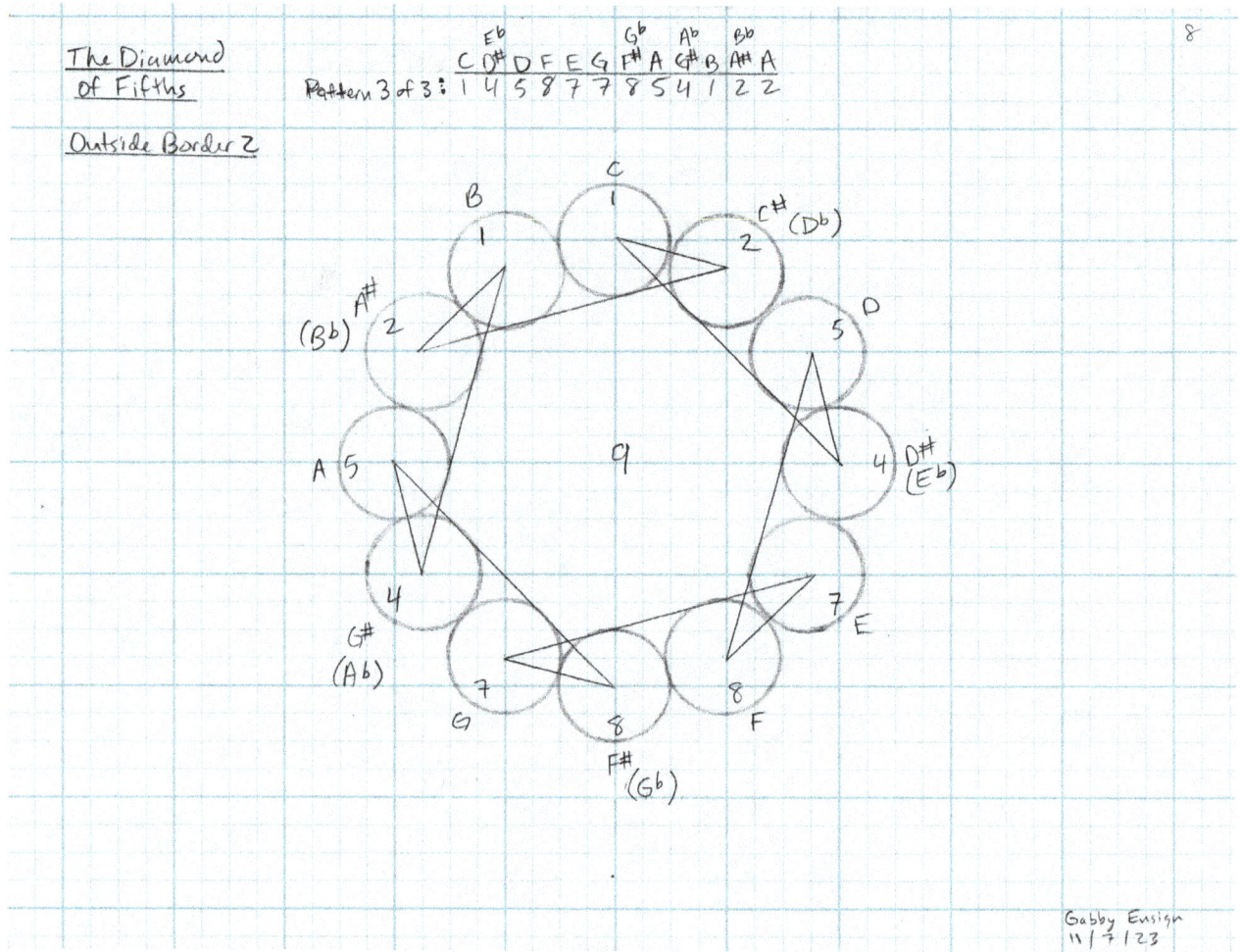

Very interesting patterns. I do not know yet their significance, but these three interlocking patterns combined show a simple geometric form which I named the Diamond of Fifths diagram. The reason why is that it follows the same pattern of the circle of fifths, and with the two outside border sequences added results in a beautiful diamond looking pattern if you were to be viewing an actual diamond from the top looking down. See the following diagram. It is the first real discovery I made leading to the previous diagram, and has my note arrangements I was postulating on the way. The dodecagon of Phi contains the Rodin 147 Shears.

I was also surprised to see the musical notes being so similar to the way 6 string guitars and five-string bass guitars were tuned. The tuning of six-string guitars tuned from the low 6th to the high 1st string is E A D G B E in three perfect intervals of fourths, followed by a major 3rd, and one more interval of a fourth. And the tuning of a five-string bass is tuned from the fifth string to the first being B E A D G; all perfect intervals of fourths. They call them fourths because with the circle of fifths chart going clockwise the intervals result in fourths and counterclockwise in fifths. You could easily name it the circle of fourths as well, but the circle of fifths is how it ended up.

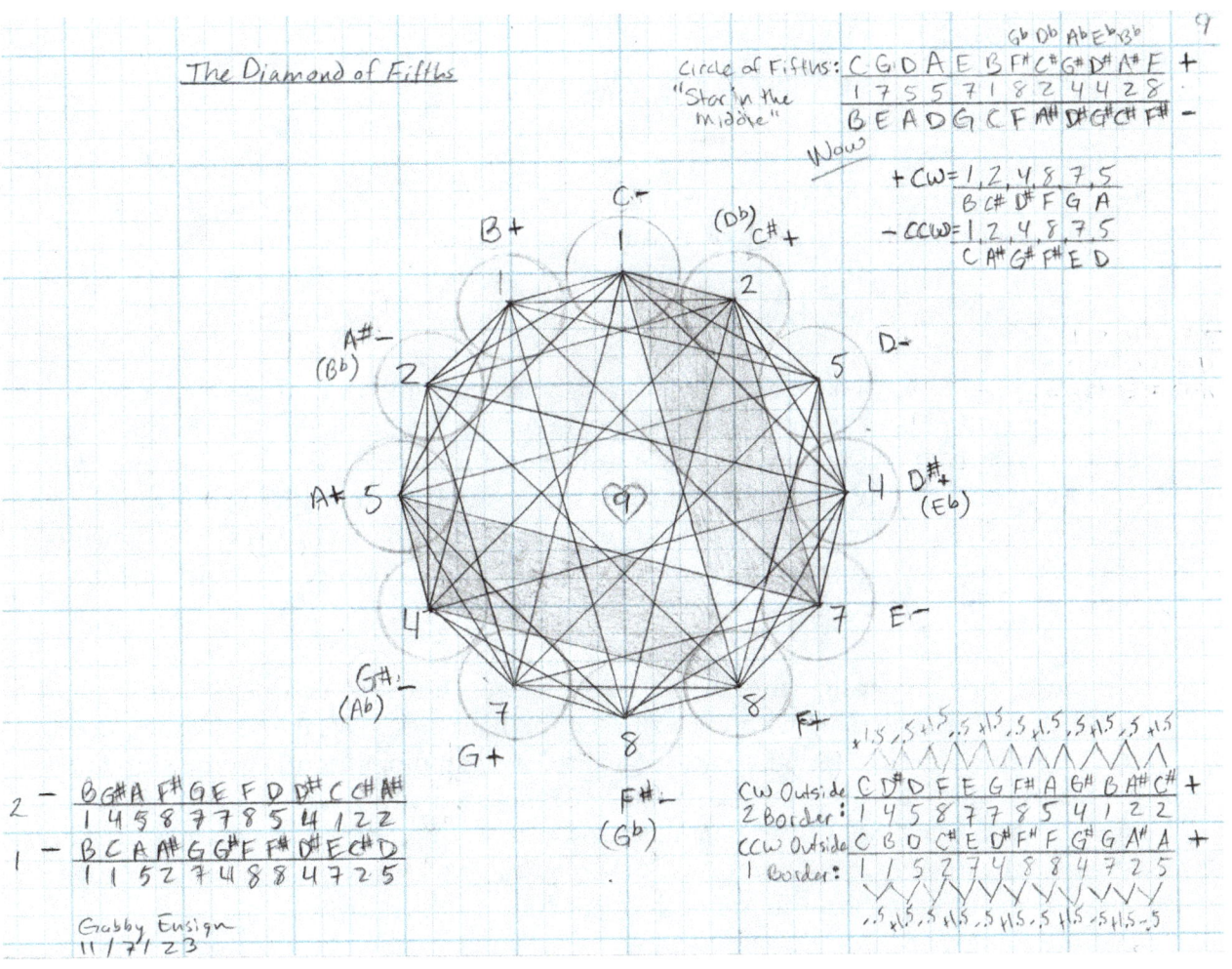

The sequence from Phi, 1 2 5 4 7 8 8 7 4 5 2 1, is a mirror of itself from the middle outward, and on the diagram, each opposite point has a digital root sum of 9. The numbers 1 + 8 = 9, 2 + 7 = 9, 5 + 4 = 9, 4 + 5 = 9, 7 + 2 = 9, and 8 + 1 = 9. You can now see the opposing polar pairs so again, we have 9 as a central point of unity within this diagram. When we look at this sequence going around the circle, it is almost exactly the same as Rodin's 147 Shears and Gator's 18-digit sequence but minus the 3 3 9 6 6 9 circuit.

Diamond of Fifths Outer Sequence
1 2 5 4 7 8 8 7 4 5 2 1
1 5 2 7 4 8 8 4 7 2 5 1
Rodin/Gator 12-digit sequence 1 of 2 (aligned)

They are the same exact sequence, except pairs of numbers are flipped, which explains the geometry of the outer border sequences of the Diamond of Fifths diagram, essentially flipping pairs of two numbers in a symmetrical way.

Phi is Vortex Based Mathematics

The relationships we have with the 18-digit sequences and everything we see within Mike Trefrei's (Gator's Donut) toroid discovery lines up perfectly with Rodin's 147 Shears. It is seen in full within Rodin's number grid maps, as well as within the split 24- digit sequence of the Fibonacci sequence I discovered. I can really appreciate why it is literally the divine proportion. And it confirms that vortex based mathematics explains its numbers in a whole new way of overlapping and counter-rotating harmonic sequences of the perfect sets of repeating numbers.

There is more regarding this in the next chapter, but this knowledge so far has established Phi's relationship with vortex based mathematics as clear and unambiguous. To me, it seemed as clear as if I could see a sunflower shining brightly in the sun when I first learned all of this. But then I also have been a student of sacred geometry for at least the last 25 years. So, I offer you a sunflower to help guide you on your path. The following drawing is an expanded Gator's Donut drawn with a compass. I hope you can see what is behind this beautiful numerical sunflower too. Enjoy!

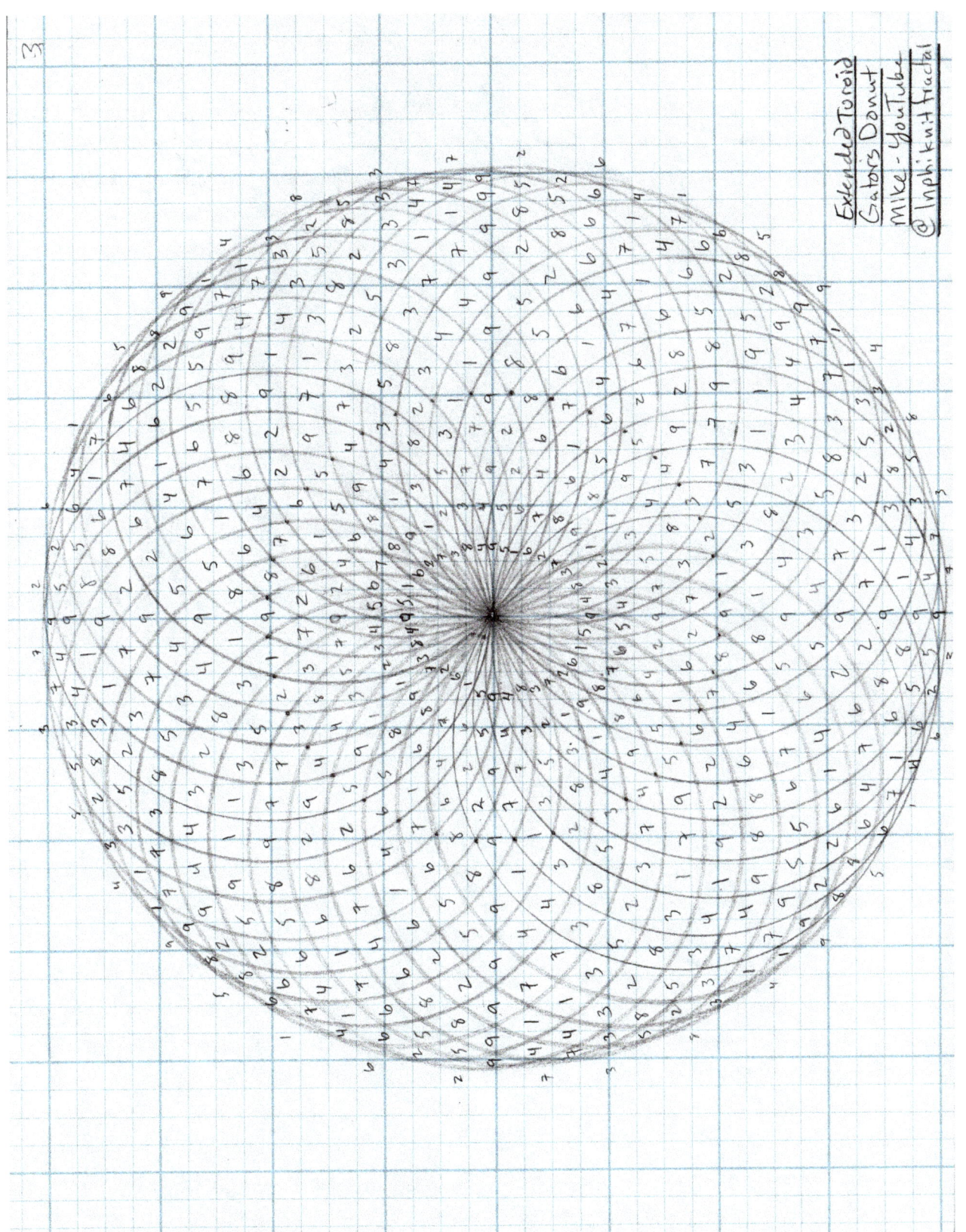

Chapter 3

The Trinity of 3 9 6 and The Divine Proportion

The Number 3

There has been a kind of intuitive sense for me that things in three are very powerful. It could have been the Catholic church's influence, but I doubt it, considering I never latched onto it through Catholic grade school, but instead kind of resisted it. Nonetheless, it contains some deeper spiritual significance to me, and it seems for many others because of its roots in so many religious triads throughout history from the Egyptian Triad to the Three Greek Fates, or the Hindu Trimurti, the Three Buddha Bodies, and in Taoism, the Three Pure Ones. I don't ascribe to any religious definition of three and creation, but if I were to describe the Holy Trinity in a spiritual, not a religious sense, I would say the Father is Creation, the Son are the Children of Creation and the Holy Spirit is the Bridge between the Father and his children in his three dimensional, seemingly separate from God, holographic illusion we call life. But if I were to describe it in vortex based mathematic terms, I would say that 9 is creation, 1 2 4 8 7 5 is spirit experiencing duality in the physical, and 3 and 6 are the bridge between creator and its creation and is the energetic driving force behind the physical world.

I don't have to go on about the number 3, because if you're reading this book, you probably already subconsciously feel it has some higher significance somehow. And if not, you can look at it in more practical terms as a function of the force that creates this universe.

In geometric terms it is an equilateral triangle of three sides with interior angles of 60° each. Put two of them together at their opposites and you have a star tetrahedron or a Star of David symbol.

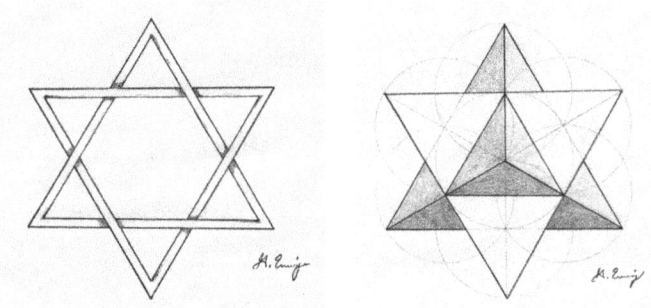

Image 23 – 24: The Star of David and the Star Tetrahedron

The Trinity of Creation

Before, when I split the 24-digit sequence of the Fibonacci ratio into two 12-digit sequences, I only showed you the one omitting the sequence with the 3 3 9 6 6 9 circuit because it operated separately from the 1 2 4 8 7 5 doubling circuit. Consider 3 9 and 6 as the unseen forces. Well, this is what the numbers showed me. See the following diagram and you'll understand why I called it what I called it. The number 9 functions at the center of two 3 9 6 triangles, one 1 1 1 triangle and one 8 8 8 triangle; in which 4 × 3 = 12, the number of points in the sequence. Since 1 and 8 are the closest to 9 in sequence is why I think they are being seen here. I do not know for sure why.

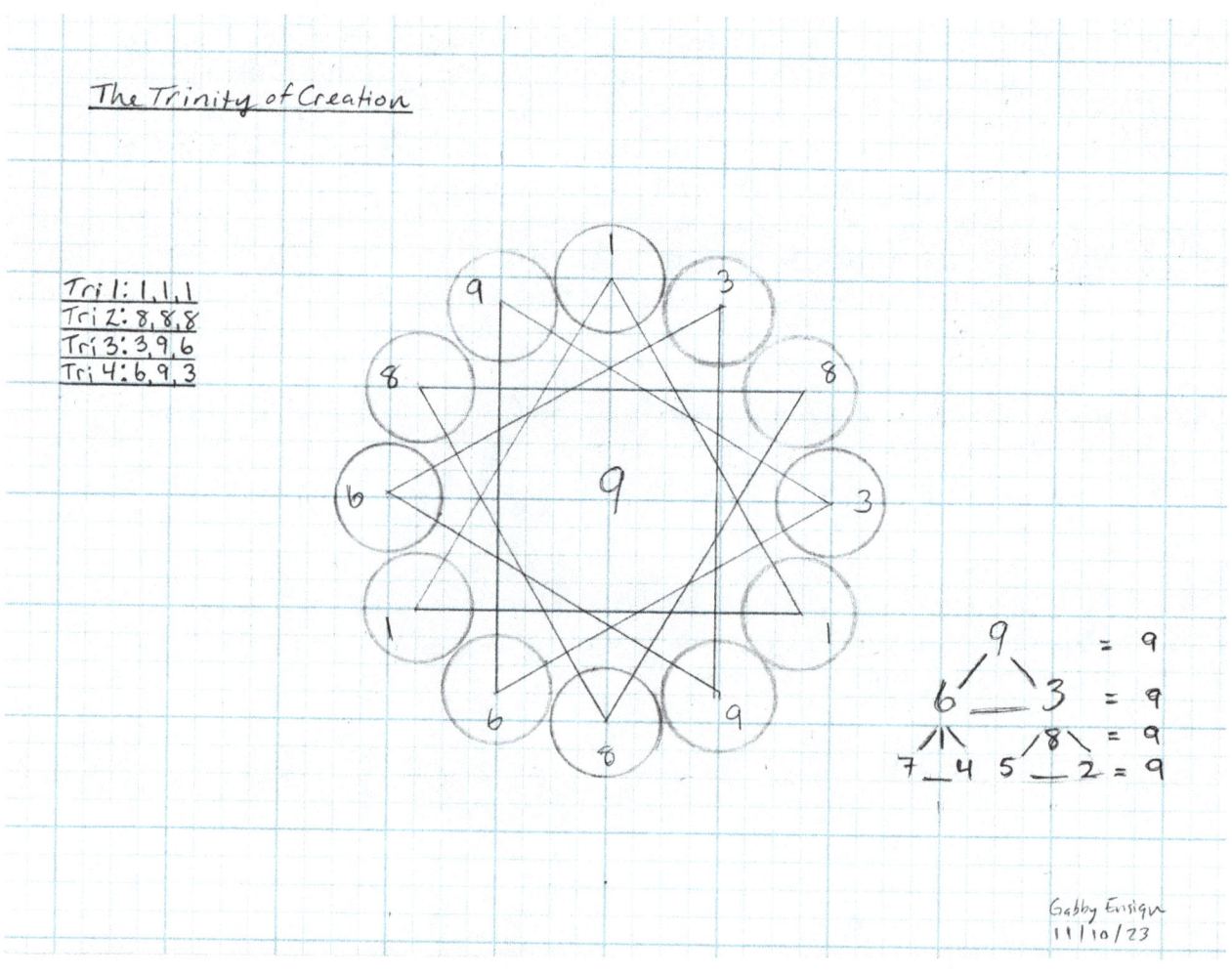

If we look at two triangles as a pair showing a star tetrahedron, then we have the 1 1 1 and 8 8 8 pairing appropriately together, and the 3 9 6 and 6 9 3 pairing appropriately. So far, this makes good sense in vortex based mathematics. Since we will soon be learning and seeing more about the 1 4 7 and 8 5 2 family number groups, take a look at the triangle diagram with the 1 and 8 as being closest in sequence to 9 and branching out to 7 4 5 and 2 on the bottom. These are merely ideas regarding why Phi chose to show only 1 and 8 as a polar pair within the opposing 12-digit sequence containing the trinity sequences of the 3 3 9 6 6 9 circuit. I would have to discuss this further with Rodin before I can understand why, but I think they may have numerical value in the sequence positioning to create the two family number groups of 1 4 7 and 8 5 2. But again, this is just a hypothesis at this point, but it is a curious thought indeed.

The next diagram shows the sequence with their hexagonal relationships, you know, the honeycomb shape. It also shows the digital root sum of all neighboring numbers, which gives us a view of the 1 2 4 8 7 5 circuit from the other side!

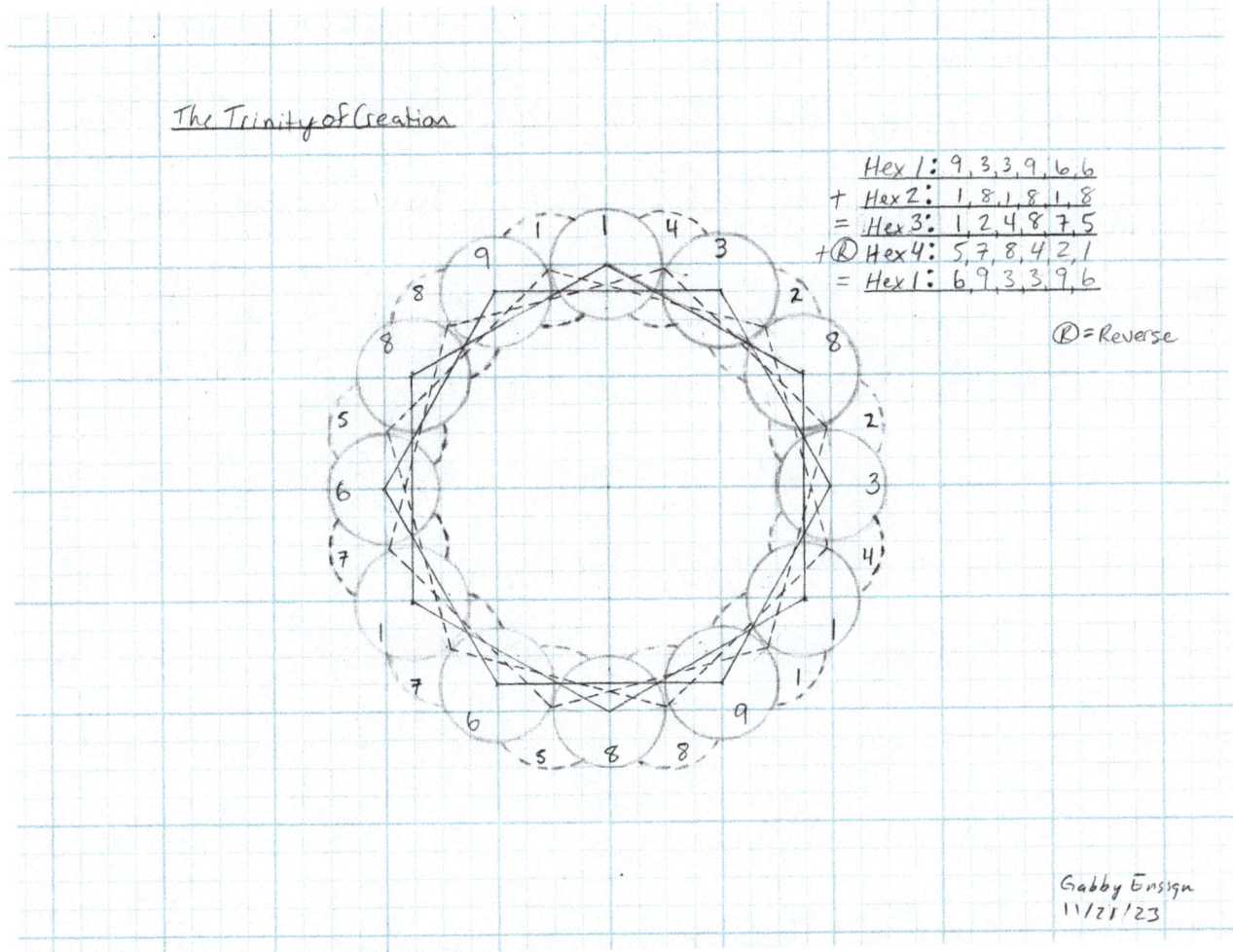

In this diagram, we have one circuit of 3 3 9 6 6 9, working with another circuit of 8 1 8 1 8 1, which coincidentally enough, if added together in their root sums, equals the 1 2 4 8 7 5 doubling circuit. You will notice the same counter-rotating sequences in the outer circles indicated with a dotted line. Through the hexagonal geometry, the six digits of the 3 3 9 6 6 9 circuit are interconnected inversely with the six digits of the doubling circuit. Very impressive indeed, and so very appropriate again, coming right off of Phi.

Holographic Root Sums

The holographic nature of the root sums of vortex based mathematics in Phi can also be seen in the next diagram showing the three 12-digit sequences at the root of both Rodin and Gator's 18-digit sequences, forming the same geometry as the Diamond of Fifths pattern seen in the first 12-digit split sequence. Again, they are inversely related to each other. The following four diagrams are each and all of the three sequences, the fifth is the combination of the geometry of the Trinity of Creation and the geometry of the Diamond of Fifths. It is mesmerizing to look at. The last is a final demonstration of the holographic effect of the first 12-digit sequence of Phi.

The Trinity of Creation
Root Sums Into
The Diamond of Fifths

Pattern 1 : 1 1 5 2 7 4 8 8 4 7 2 5
Pattern 2 : 1 4 5 8 7 7 8 5 4 1 2 2
Pattern 3 : 1 7 5 5 7 1 8 2 4 4 2 8

In the beginning
was the word...
...and the Word was Love,
and Love became Sound,
and Sound became Harmony,
and Harmony became Music,
and Music became Geometry,
and Geometry became Form,
and Form became the
vessel to house the
Son of God, in his
experience of Form.

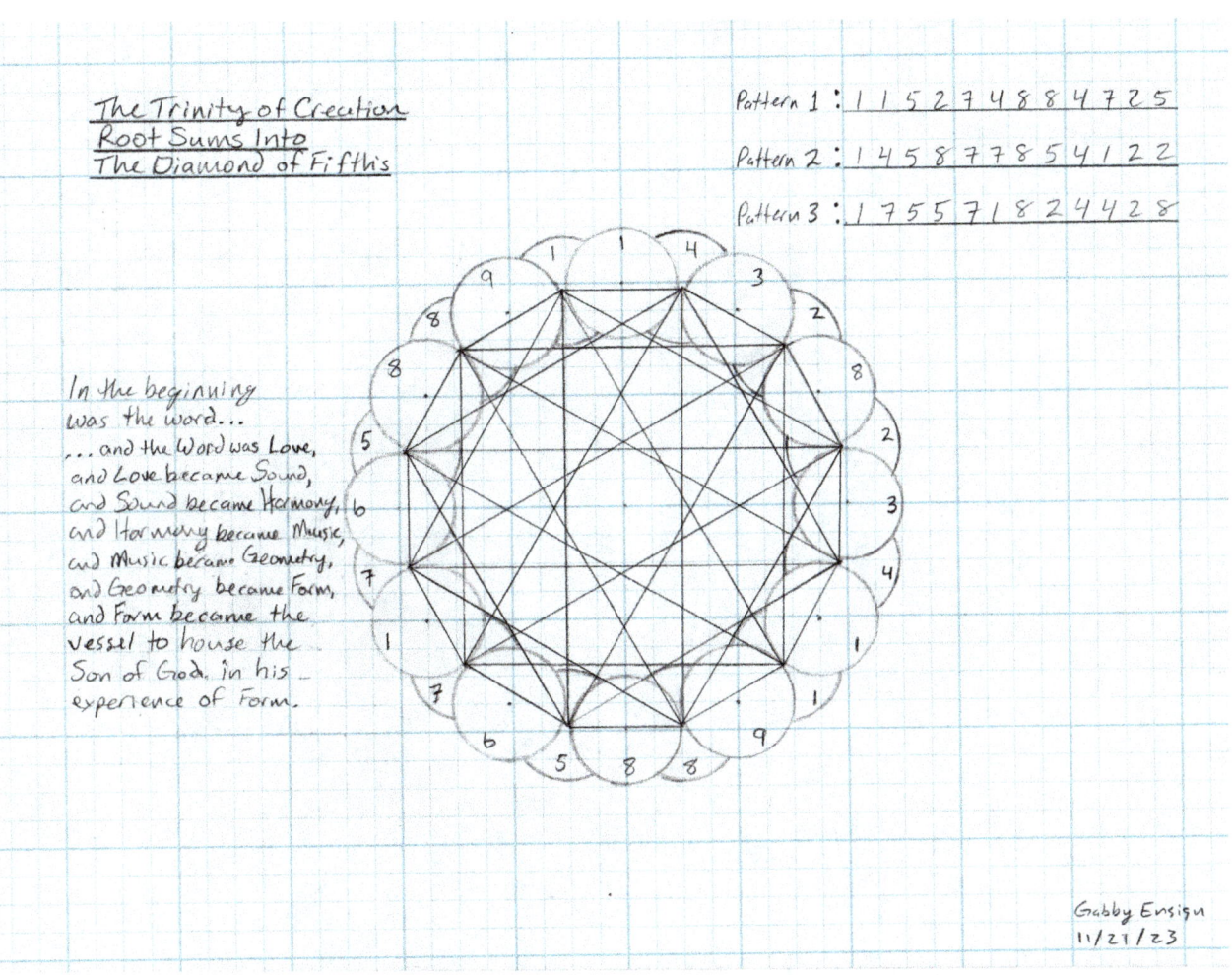

Gabby Ensign
11/27/23

<u>The Trinity of Creation</u>
<u>Root Sums Into</u>
<u>The Diamond of Fifth's</u>

Pattern 1: 1 1 5 2 7 4 8 8 4 7 2 5

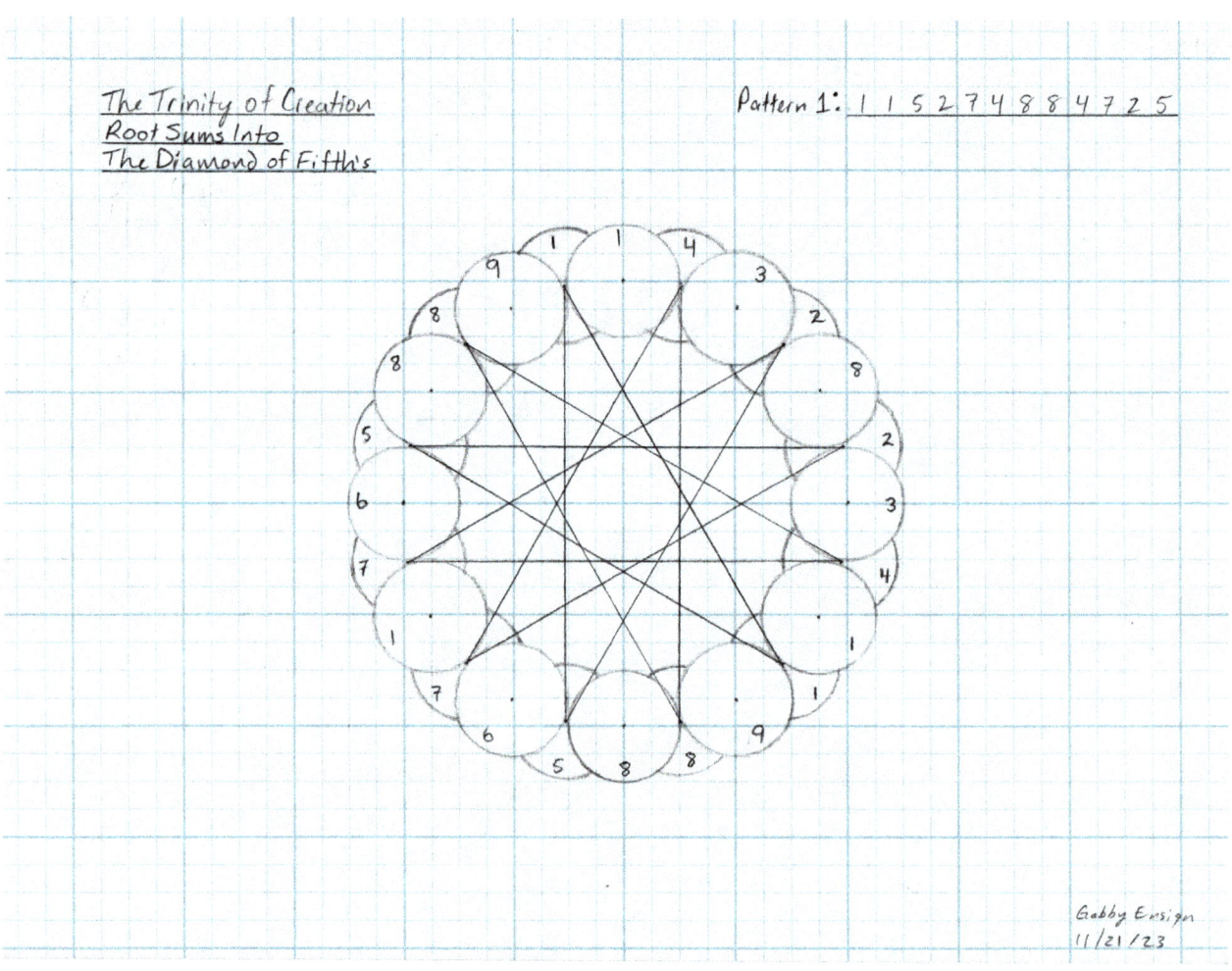

Gabby Ensign
11/21/23

The Trinity of Creation
Root Sums Into
The Diamond of Fifth's

Pattern 2: 1 4 5 8 7 7 8 5 4 1 2 2

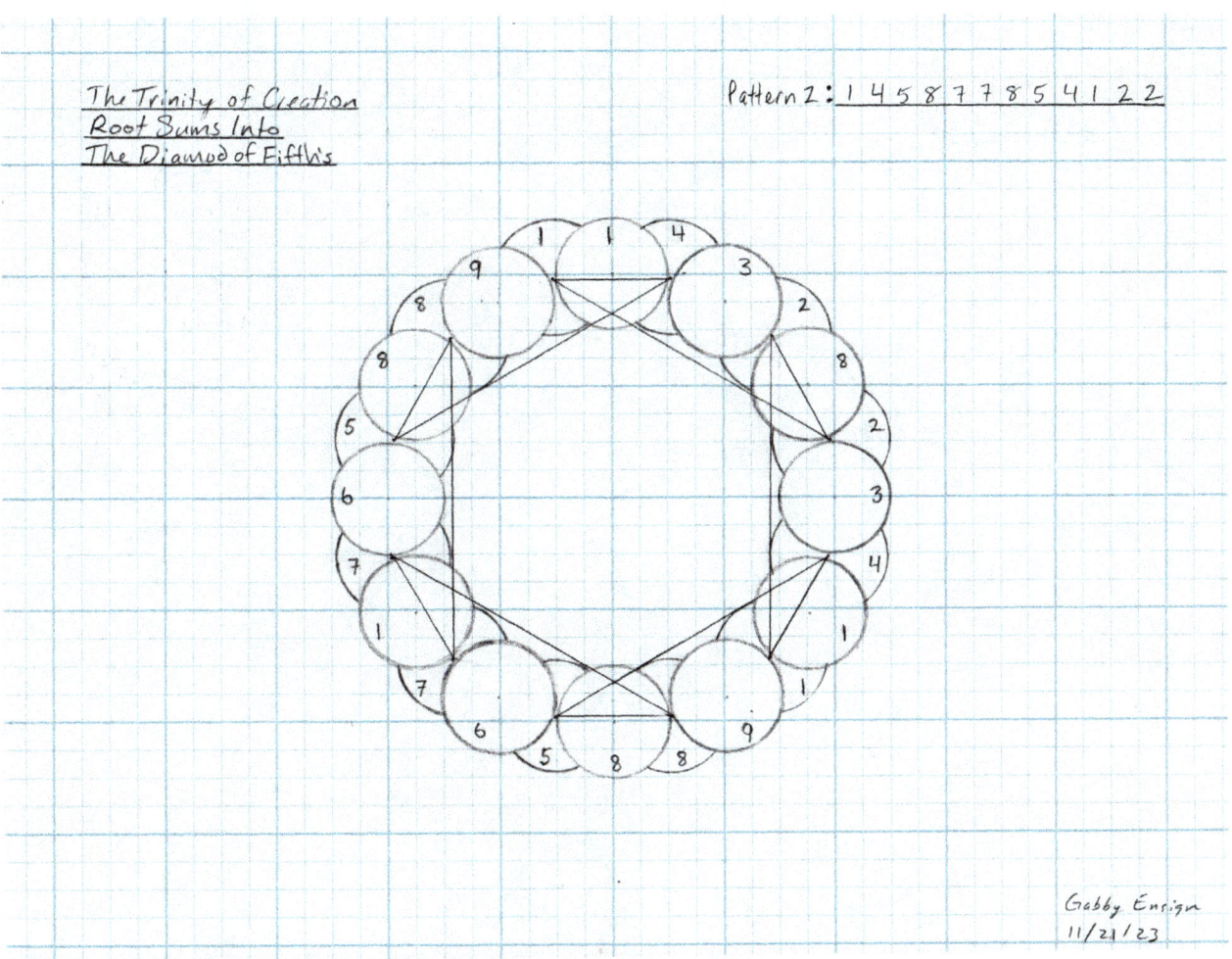

Gabby Ensign
11/21/23

The Trinity of Creation
Root Sums Into
The Diamond of Fifth's

Pattern 3: 1 7 5 5 7 1 8 2 4 4 2 8

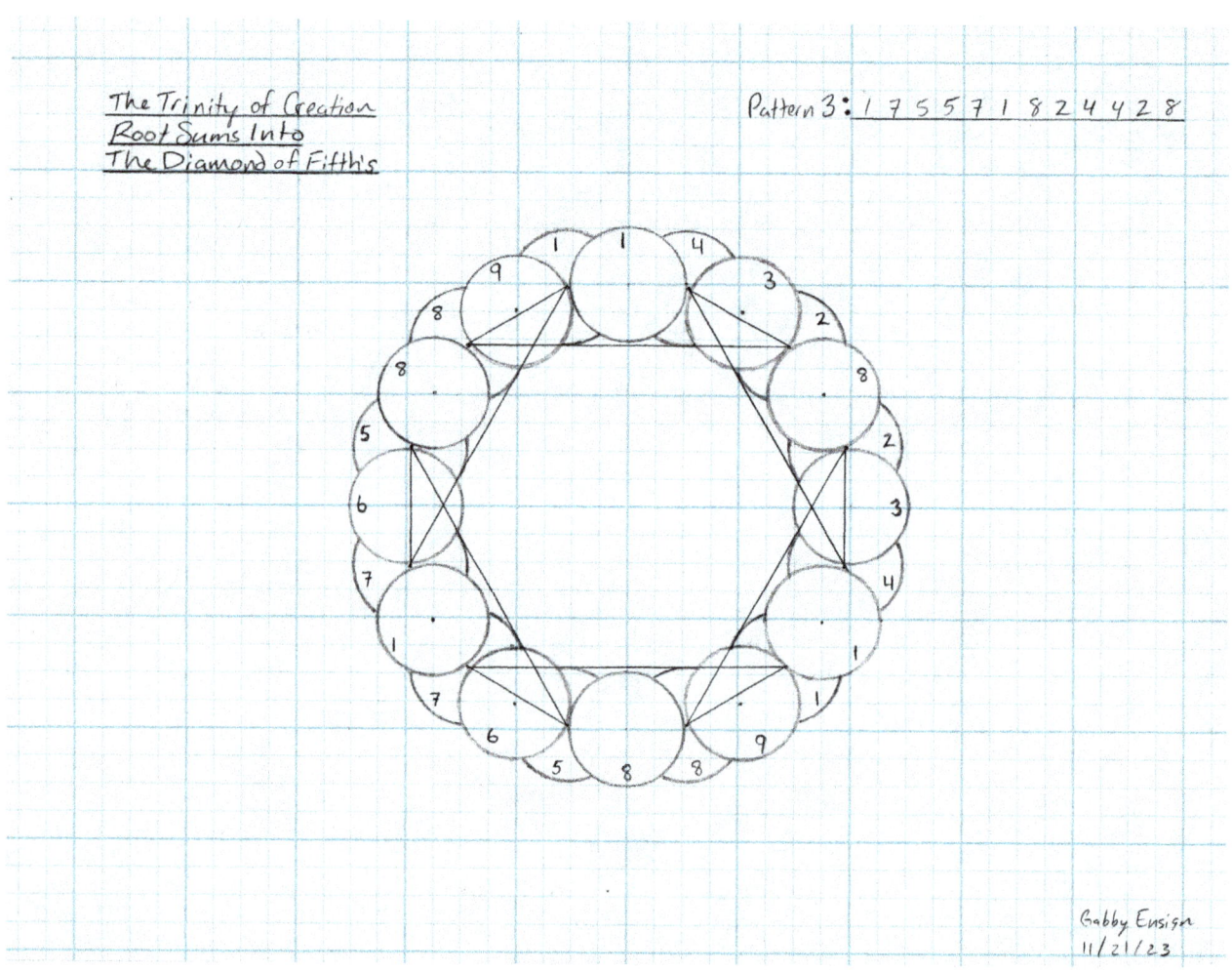

Gabby Ensign
11/21/23

The Trinity of Creation and The Diamond of Fifth's

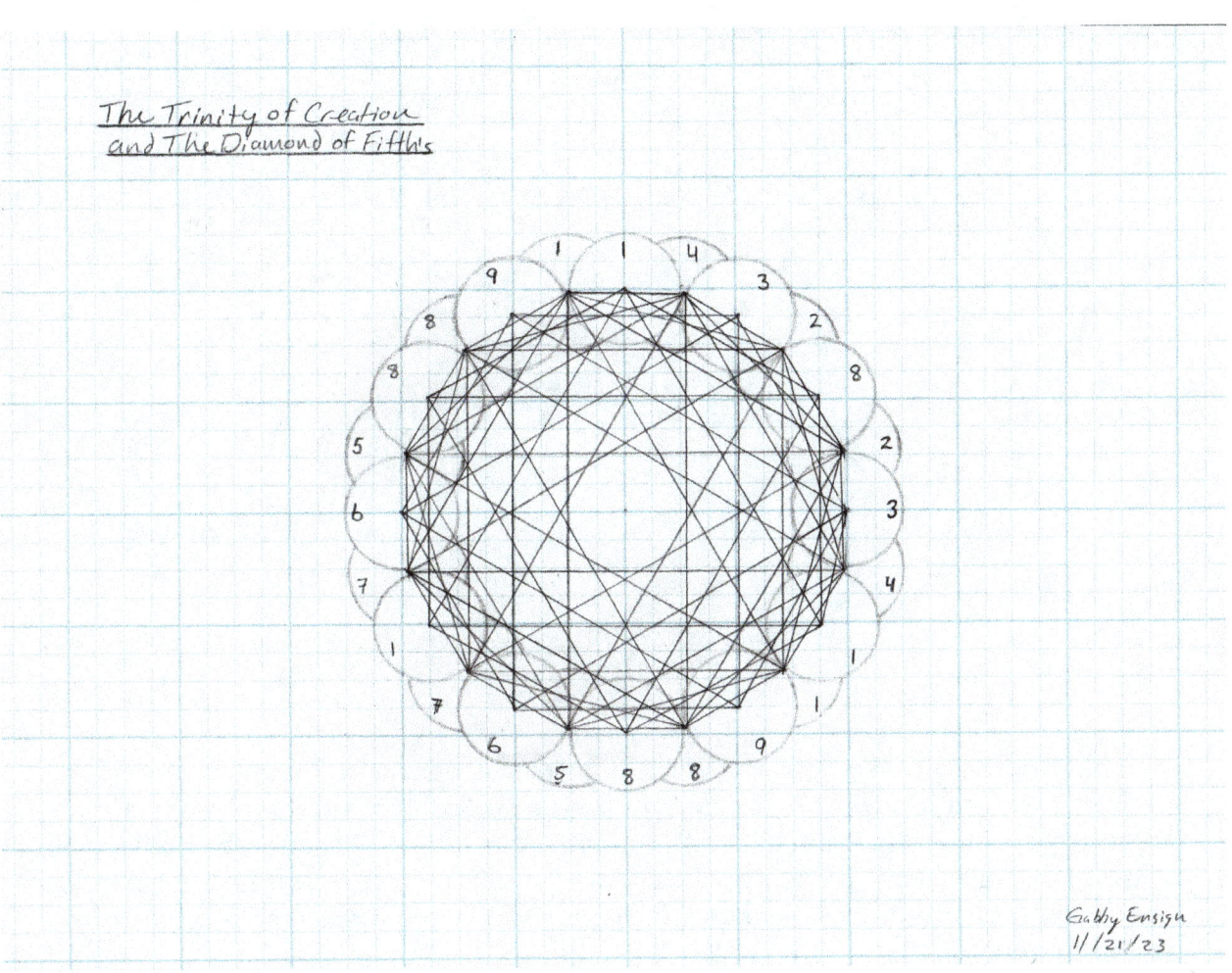

Gabby Ensign
11/21/23

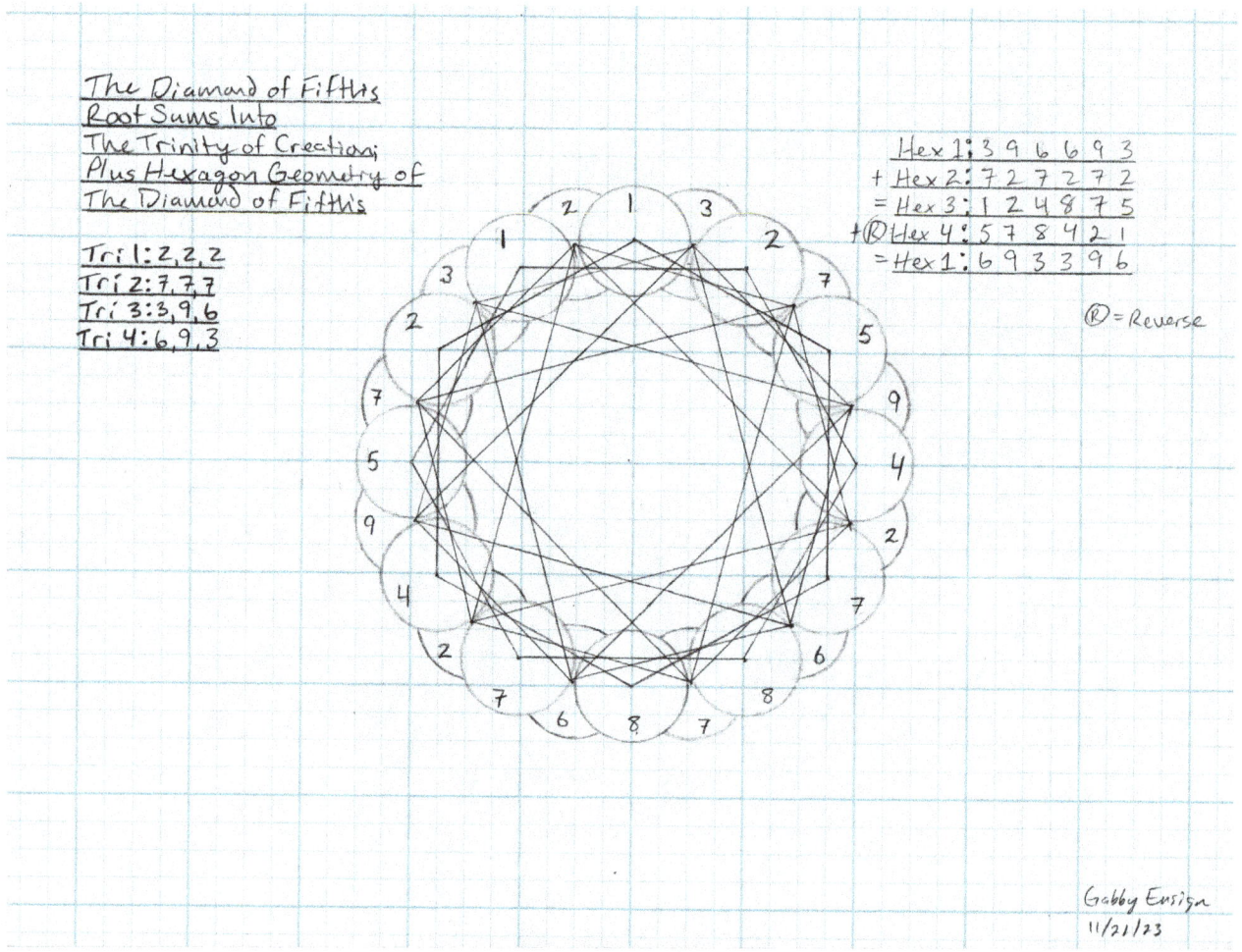

This type of interlaced connectedness with the sequences and numbers is because of their harmonic nature and the rules in their relationships with each other. This is not in any personal sense, but in a programming sense. Each sequence does tell a type of story through the expression of their geometry and their place within the structure of the numbers. It is even more incredible that vortex based mathematics correlates so intimately with Phi.

There really is immense information on this magical and mystical number of Phi. I highly recommend two books written by a mathematician named Mario Livio called, *"The Golden Ratio – The Story of Phi, The World's Most Astonishing Number,"* and *"Is God a Mathematician?"*

The Phi Flower of Sums

Well, it is time to look behind another flower I discovered, this one a diagram of the 24-digit sequence of Phi and the root sum of every three numbers, the counter-rotating primary and secondary circuits that are present, and their root sum of every third number and their primary and secondary circuits outward to three extensions. It works in threes I tell you, for all the polar pairs are there.

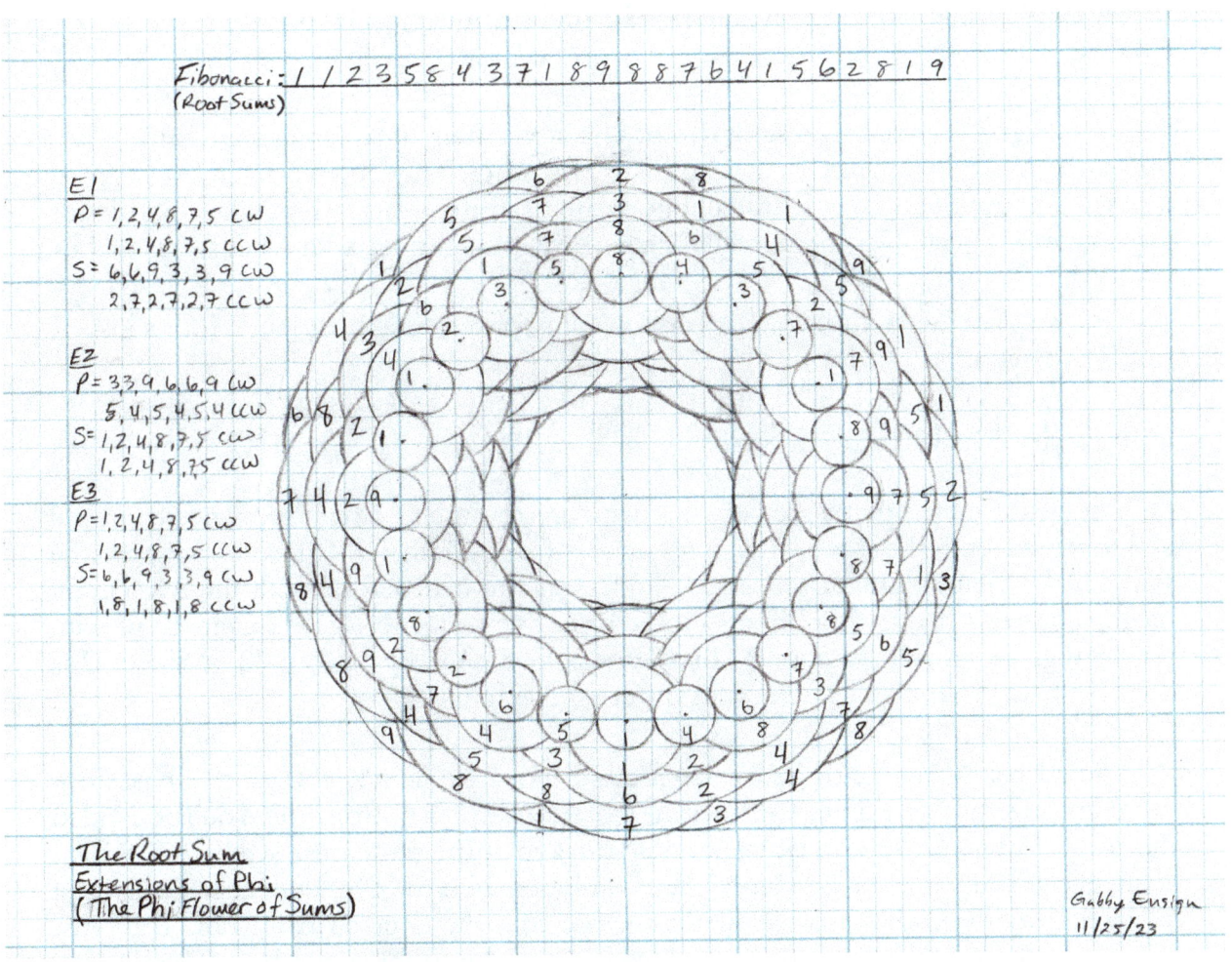

Fibonacci: 1 1 2 3 5 8 4 3 7 1 8 9 8 8 7 6 4 1 5 6 2 8 1 9
(Root Sums)

E1
P = 1,2,4,8,7,5 CW
    1,2,4,8,7,5 CCW
S = 6,6,9,3,3,9 CW
    2,7,2,7,2,7 CCW

E2
P = 3,3,9,6,6,9 CW
    5,4,5,4,5,4 CCW
S = 1,2,4,8,7,5 CW
    1,2,4,8,7,5 CCW

E3
P = 1,2,4,8,7,5 CW
    1,2,4,8,7,5 CCW
S = 6,6,9,3,3,9 CW
    1,8,1,8,1,8 CCW

The Root Sum
Extensions of Phi
(The Phi Flower of Sums)

Gabby Ensign
11/25/23

If you look at the outer larger circles, extending outward from the smaller inner circles, each circle in front is the primary, and each circle in the back is the secondary. E1 is the most inner set of circles representing the first extension, E2 the second, and E3 is the third.

In E1, the two primary circuits counter-rotating with each other are 1 2 4 8 7 5 doubling circuits. The secondary has one 3 3 9 6 6 9 circuit, and a counter-rotating 7 2 7 2 7 2 circuit. E2 in its primary has one 3 3 9 6 6 9 circuit and one 5 4 5 4 5 4 circuit, and in its secondary two counter-rotating 1 2 4 8 7 5 doubling circuits again. Needless to say, E3 flips back to a primary 1 2 4 8 7 5 and secondary 3 3 9 6 6 9 and 8 1 8 1 8 1 circuits.

The diagram is very representative of the active numerical circuits within Phi and the numerical functionality that it contains. Its inherent geometry fractals out in its design. The circles are merely visual, but create beautifully layered mandala like clouds looking into heaven. Geometry really is a beautiful thing, no matter what form it takes.

I am so grateful to *"@WhyPhi?"* on YouTube for teaching people the root sums of Phi. Thank you.

The Rodin Phi Flower of Sums

The next four diagrams are very fascinating. I took the 24-digit sequence from Phi and merged it with the three 18-digit sequences from Rodin's 147 Shears, which are the same as the Gator's Donut sequences, and as mentioned before, with the 396 circuit having shifted. I had to run Phi into three repeating sequences of 72 digits to fit with the 18-digit sequences because 24 or 48 are not divisible by 18, but 72 is. Since 72 is divisible by 18 four times we get four different diagrams.

Once I placed the 72 digits of Phi around the circle, I took the root sum of four numbers together at a time, considering that "@WhyPhi?" from the YouTube video did three. I wanted to see what four would do, and at the same time see if the 18-digit sequence would show me anything when combined with the Phi sequence. The largest 18 circles are the root sums encompassing the four digits taken from Phi. Then just outside of these circles is the first extended layer, split into three numbers. They are the three 18-digit sequences side-by-side as they wrap around clockwise. Just beyond that is another extended layer that is split into three numbers. They are the root sum of Phi's sum of four numbers combined with each number of the 18-digit sequences. Now that you understand how I got the numbers, let's talk about what I discovered.

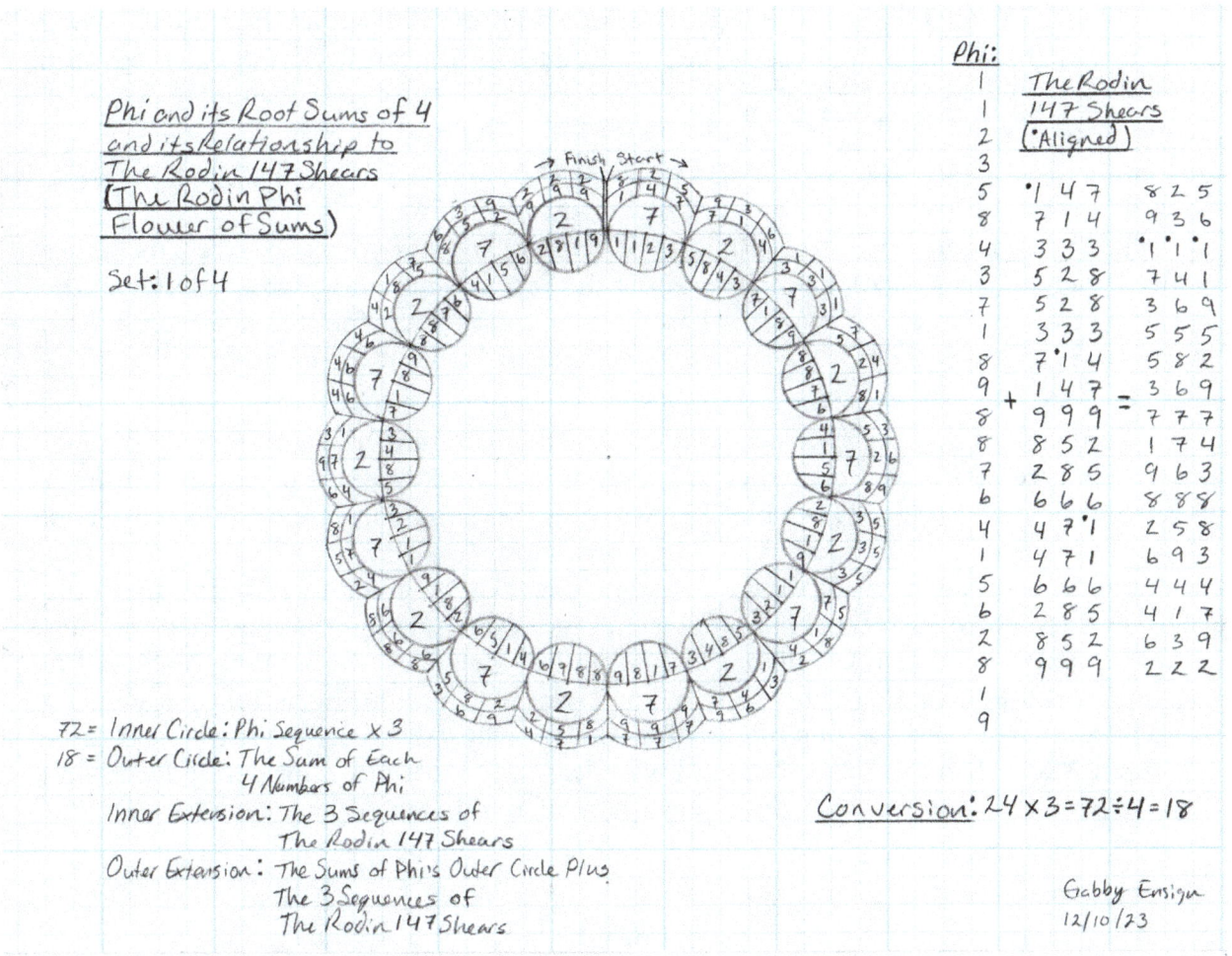

In the first diagram, we get a series of 7 and 2 repeating around the circle of Phi. The first set of sums of four are 1 + 1 + 2 + 3 = 7, then 5 + 8 + 4 + 3 = 20 and 2 + 0 = 2 and so on. We know that there are four polar pairs, 1 and 8, 2 and 7, 3 and 6, and 4 and 5. At this point one might assume that the other three diagrams will give us the other three sets. I did and I was wrong. They were something else which you will see in just a moment. Moving on, starting at the 90° position of where Phi begins with its first sum of 7, we have 147. The 1 on the left begins the sequence, 1 7 3 5 5 3 7 1 9 8 2 6 4 4 6 2 8 9 going clockwise around the full circle. The 4 in the middle begins the sequence 4 1 3 2 2 3 1 4 9 5 8 6 7 7 6 8 5 9, and 7 on the right with 7 4 3 8 8 3 4 7 9 2 5 6 1 1 6 5 2 9. The reason you might not recognize the sequences is because they are usually written starting at 1. I have aligned them to their inherent family number groups of 1 4 7, 8 2 5 and 3 9 6, and they are indicated with a ● symbol at the beginning of each sequence at 1 in the diagram.

Now that we have established the root sums of Phi within four neighboring digits and we know the three sequences of Rodin's 147 Shears, let's add them together. Starting with the first 7, we have 7 + 1 = 8, 7 + 4 = 11 = 1 + 1 = 2, and 7 + 7 = 14 = 1 + 4 = 5. Therefore, 147 gets you 825 with Phi. Fabulous!

Harmony + harmony = harmony, obviously. Let's move onto the next circle; 2 + 7 = 9, 2 + 1 = 3, and 2 + 4 = 6; 714 gets 936 from 2; another vortex based mathematics fundamental. At this point I already know the rest will fall into place showing all the symmetry I expect to see in vortex based mathematics. My experiment worked. Next, we see 111 from 7 and 333, 741 from 2 and 528, and around the circle we go!

Something I noticed right away was that the three combined numbers from the first layer that are all the same are either 333, 666, or 999. And the sums on the outside layer have triple digit numbers of 111, 222, 444, 888, 777, and 555. 1 2 4 8 7 5. Very cool indeed. There is some definite flip-flopping going on again, just as vortex based mathematics is expected to show us. Looking at them as a whole, they are the Rodin 147 Shears, except that the 3 3 9 6 6 9 circuit has shifted within the sequence and they are aligned to number 1. Go figure. The Rodin 147 Shears + 4 Root Sums (4RS) of Phi = the Rodin 147 Shears (with a shifted 3 3 9 6 6 9 circuit). Shifting and flipping. This is a twisted science indeed, pun intended. I could hardly wait to see what the next three diagrams would show me. Let me say, I was pleasantly surprised, considering I was kind of expecting to see the other three polar pairs of numbers. Nope… even better.

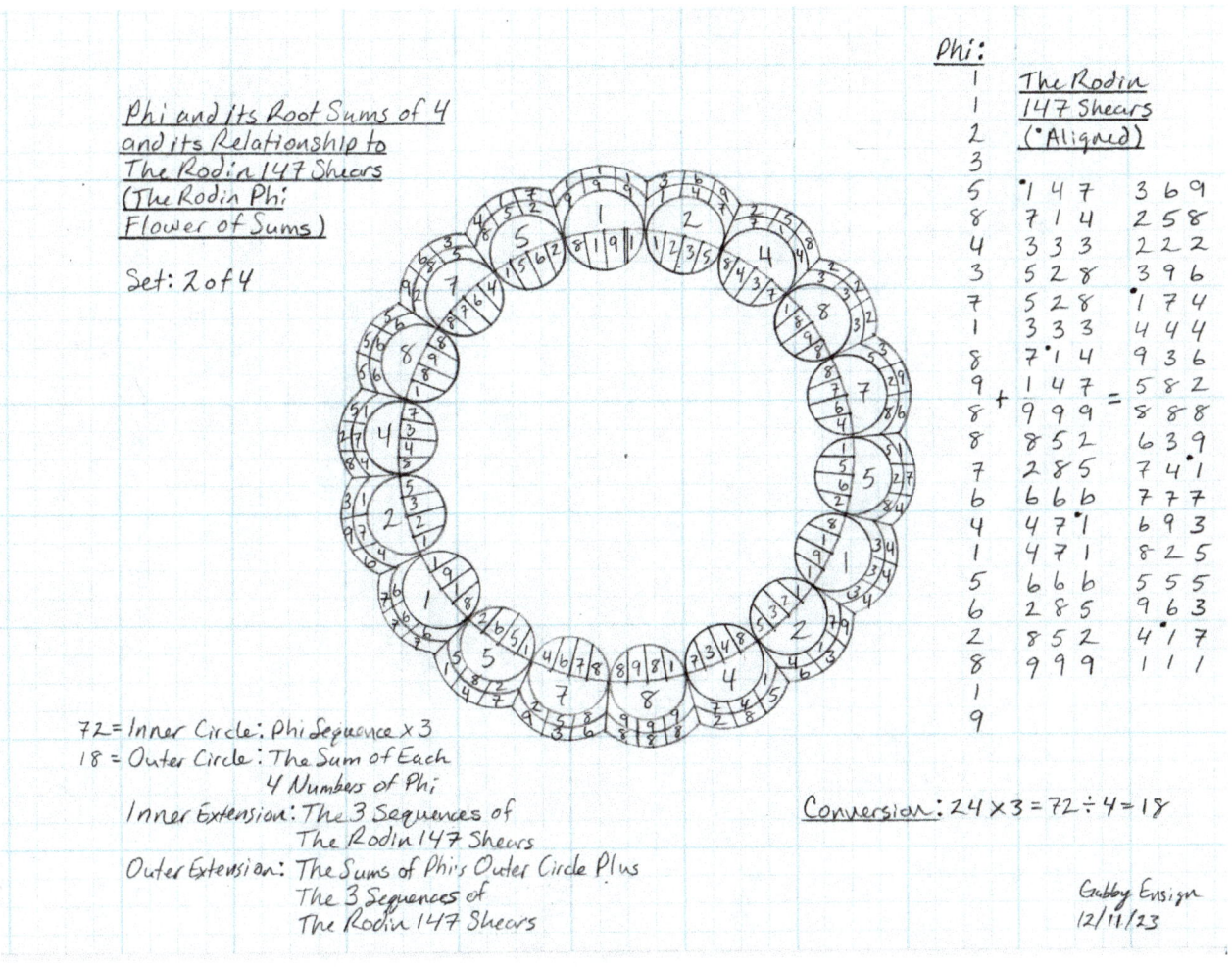

Phi and its Root Sums of 4 and its Relationship to The Rodin 147 Shears (The Rodin Phi Flower of Sums)

Set: 2 of 4

72 = Inner Circle: Phi Sequence x 3
18 = Outer Circle: The Sum of Each 4 Numbers of Phi
Inner Extension: The 3 Sequences of The Rodin 147 Shears
Outer Extension: The Sums of Phi's Outer Circle Plus The 3 Sequences of The Rodin 147 Shears

Phi:	The Rodin 147 Shears (*Aligned)
1
1
2
3
5	1 4 7 3 6 9
8	7 1 4 2 5 8
4	3 3 3 2 2 2
3	5 2 8 3 9 6
7	5 2 8 1 7 4
1	3 3 3 4 4 4
8	7 1 4 9 3 6
9	+ 1 4 7 = 5 8 2
8	9 9 9 8 8 8
8	8 5 2 6 3 9
7	2 8 5 7 4 1
6	6 6 6 7 7 7
4	4 7 1 6 9 3
1	4 7 1 8 2 5
5	6 6 6 5 5 5
6	2 8 5 9 6 3
2	8 5 2 4 1 7
8	9 9 9 1 1 1
1
9

Conversion: 24 x 3 = 72 ÷ 4 = 18

Gabby Ensign
12/11/23

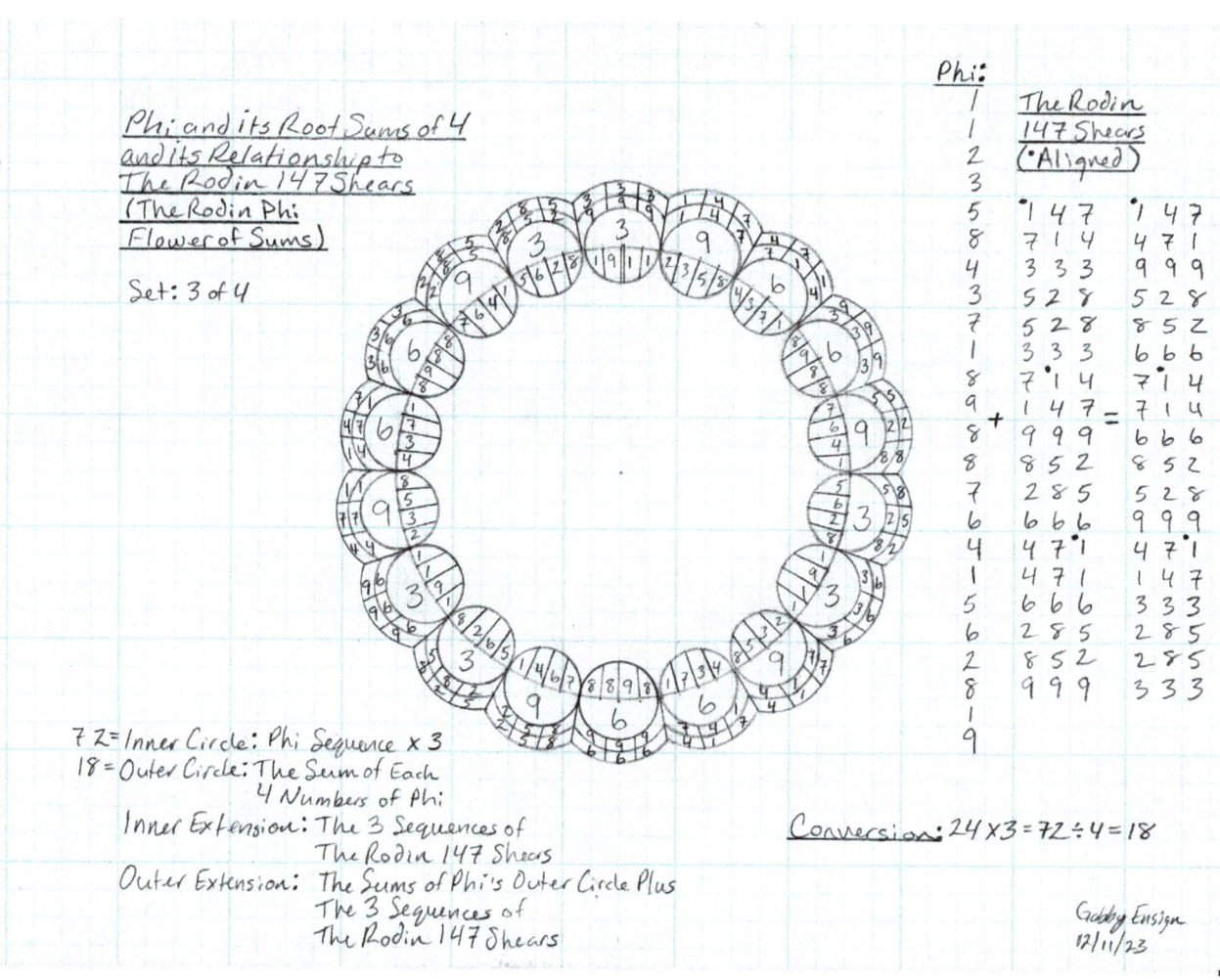

Phi and its Root Sums of 4 and its Relationship to The Rodin 147 Shears
(The Rodin Phi Flower of Sums)

Set: 3 of 4

72 = Inner Circle: Phi Sequence × 3
18 = Outer Circle: The Sum of Each 4 Numbers of Phi
Inner Extension: The 3 Sequences of The Rodin 147 Shears
Outer Extension: The Sums of Phi's Outer Circle Plus The 3 Sequences of The Rodin 147 Shears

Phi:		The Rodin 147 Shears (•Aligned)	
1			
1			
2			
3			
5		•1 4 7	•1 4 7
8		7 1 4	4 7 1
4		3 3 3	9 9 9
3		5 2 8	5 2 8
7		5 2 8	8 5 2
1	+	3 3 3	= 6 6 6
8		•7 1 4	•7 1 4
9		1 4 7	7 1 4
8		9 9 9	6 6 6
8		8 5 2	8 5 2
7		2 8 5	5 2 8
6		6 6 6	9 9 9
4		•4 7 1	•4 7 1
1		4 7 1	1 4 7
5		6 6 6	3 3 3
6		2 8 5	2 8 5
2		8 5 2	2 8 5
8		9 9 9	3 3 3
1			
9			

Conversion: 24 × 3 = 72 ÷ 4 = 18

Gabby Ensign
12/11/23

Phi and its Root Sums of 4 and its Relationship to The Rodin 147 Shears (The Rodin Phi Flower of Sums)

Set: 4 of 4

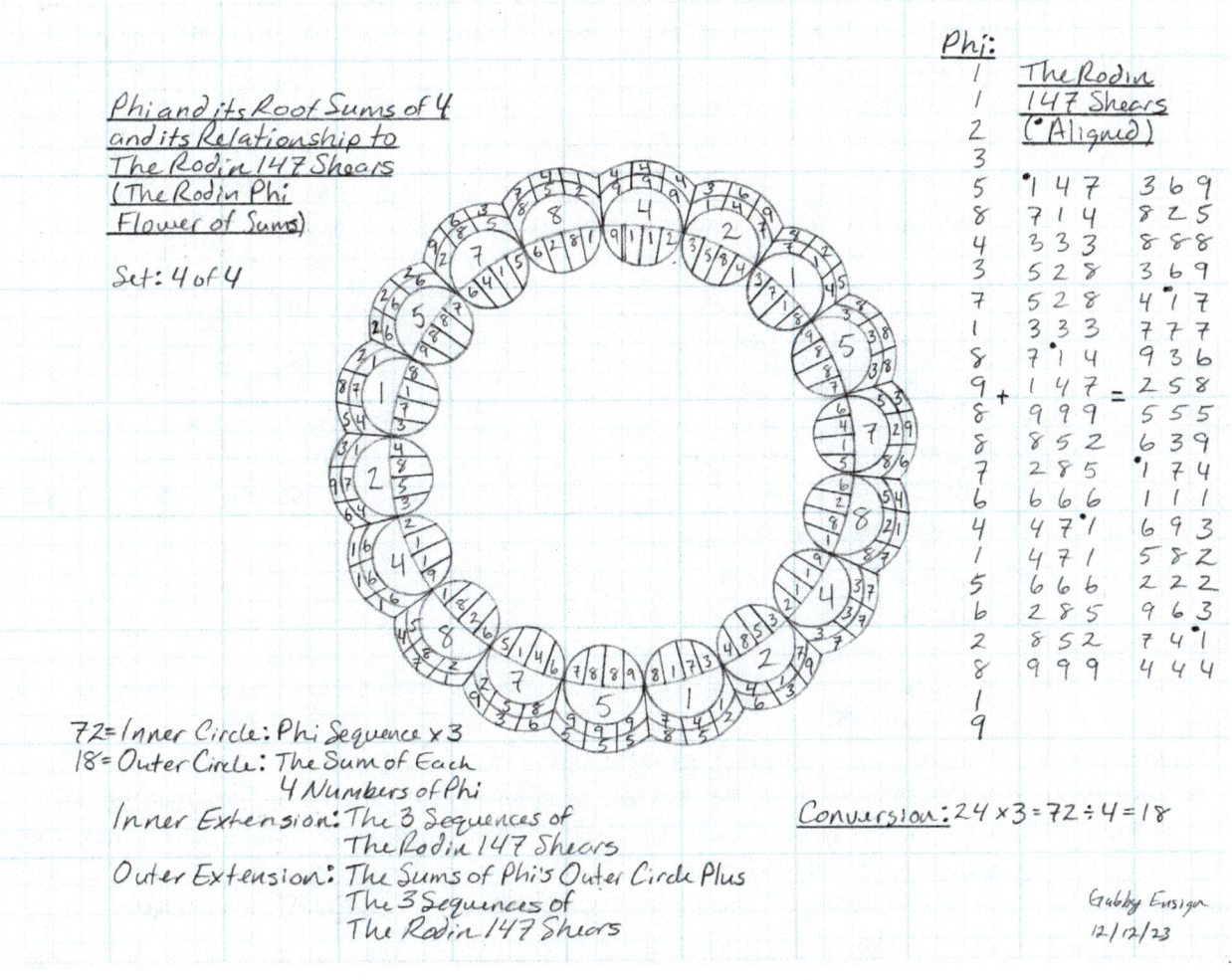

72 = Inner Circle: Phi Sequence x 3
18 = Outer Circle: The Sum of Each 4 Numbers of Phi
Inner Extension: The 3 Sequences of The Rodin 147 Shears
Outer Extension: The Sums of Phi's Outer Circle Plus The 3 Sequences of The Rodin 147 Shears

Phi:

Phi		The Rodin 147 Shears (*Aligned)	
1			
1			
2			
3			
5	*	1 4 7	3 6 9
8		7 1 4	8 2 5
4		3 3 3	8 8 8
3		5 2 8	3 6 9
7		5 2 8	4 1 7
1		3 3 3	7 7 7
8		7 1*4	9 3 6
9	+	1 4 7	= 2 5 8
8		9 9 9	5 5 5
8		8 5 2	6 3 9
7		2 8 5	*1 7 4
6		6 6 6	1 1 1
4		4 7*1	6 9 3
1		4 7 1	5 8 2
5		6 6 6	2 2 2
6		2 8 5	9 6 3
2		8 5 2	7 4 1
8		9 9 9	4 4 4
1			
9			

Conversion: 24 x 3 = 72 ÷ 4 = 18

Gabby Ensign
12/12/23

How wonderful is that? From Phi, depending on which four neighboring numbers we add together, gets us the doubling circuit of 1 2 4 8 7 5, the 3 3 9 6 6 9 circuit and the polar pair of 2 and 7. The doubling circuit is seen twice with the second one going in the opposite direction. There is a reciprocal back-and-forth. I am very fond of these diagrams because anything to do with Phi I find fascinating. The fact that 1 2 4 8 7 5 and 3 3 9 6 6 9 were so apparent makes Phi that much more fascinating. We saw it in @*WhyPhi?*'s diagram and are supercharged in these diagrams. They are even kind of pretty to look at, like another flower in the garden of sunflowers. I'm not sure why we see 2 and 7 and no other polar pairs, but maybe someday we will know. We have seen only 1 and 8 elsewhere.

I also do not know how this diagram serves its function within the framework of creation, but it is showing us the nature of harmonic numbers and their harmonic outcomes when combined with other harmonic numbers.

I do know that it is further validation of the inherent nature of vortex based mathematics within Phi. Since Phi is basically an expression of everything, vortex based mathematics is the key to looking at the functions behind Phi. It's as if you are seeing the machine as Phi and the ghost as vortex based mathematics. Therefore, vortex based mathematics explains the ghost in the machine if you will. No, I'm serious, if 9 is spirit, and 3 and 6 are the bridge, 1 2 4 8 7 5 are the parts, and Phi is the assemblance of those parts, then without spirit, the rest is empty.

Holographic Sequences

I would like to wrap this chapter up with these last set of diagrams because their sequences are so apparent in vortex based mathematics. They also tied together the geometry of Phi, so their geometry must play an important role in our universe. That being said, I would like to add that there are many geometries, all playing their part in the creation process, but since these three sequences are seen extensively in vortex based mathematics, I attempted to see how they related to each other in geometric sequence. Sure enough, they were all holographic of each other and themselves.

The 12-digit sequence of Phi creating the Diamond of Fifths diagram is essentially a dodecagon which is a 12 sided polygon. If you were to take a point and draw a line to each other point, you would have something similar to the Diamond of Fifths or circle of fifths, but it wouldn't be the same. The lines from point to point skip one in between, meaning not all points are supposed to connect. These patterns of geometry are harmonious and are holographically expressing identical geometries. The geometry of Rodin and Gator (minus 3 3 9 6 9 9), Phi and the Diamond of Fifths all incorporate a dodecagon.

Rodin's 1, 4 & 7 Shears
Sequences 1, 2 & 3 Combined*

Sequence 1:
○ 1 1 5 2 7 4 8 8 4 7 2 5

Sequence 2:
1 4 5 8 7 7 8 5 4 1 2 2

Sequence 3:
1 7 5 5 7 1 8 2 4 4 2 8

*18 digit sequence minus 6 digit 3, 3, 9, 6, 6, 9 sequence equals a 12 digit sequence.

○ Indicates outer sequence

Clockwise ----:
1 2 4 8 7 5

Counterclockwise ----:
1 2 4 8 7 5

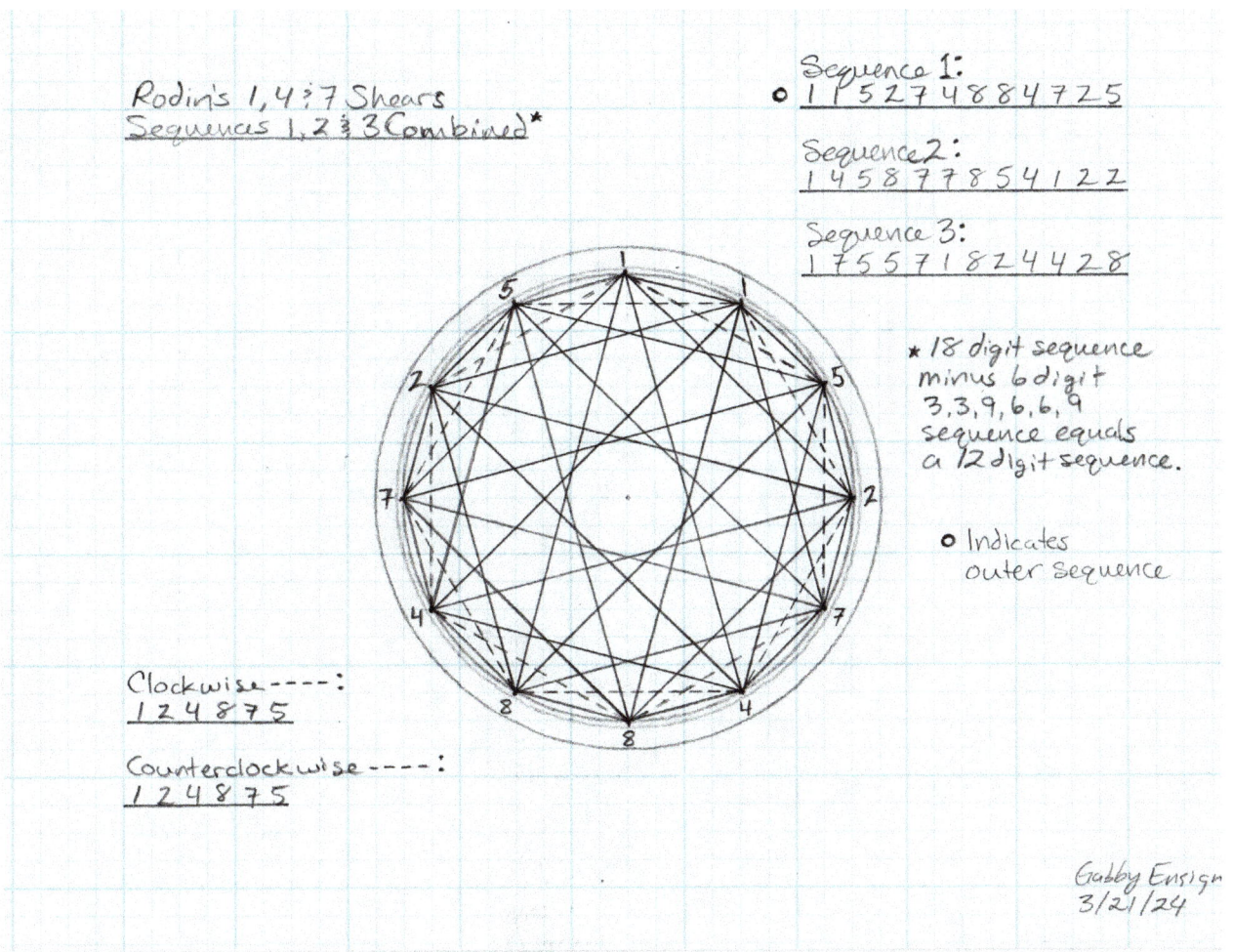

Gabby Ensign
3/21/24

Rodin's 1, 4 & 7 Shears
Sequences 1, 2 & 3 Combined*

Sequence 1:
115274884725

Sequence 2:
○ 145877854122

Sequence 3:
175571824428

* 18 digit sequence minus 6 digit 3,3,9,6,6,9 sequence equals a 12 digit sequence.

○ Indicates outer sequence

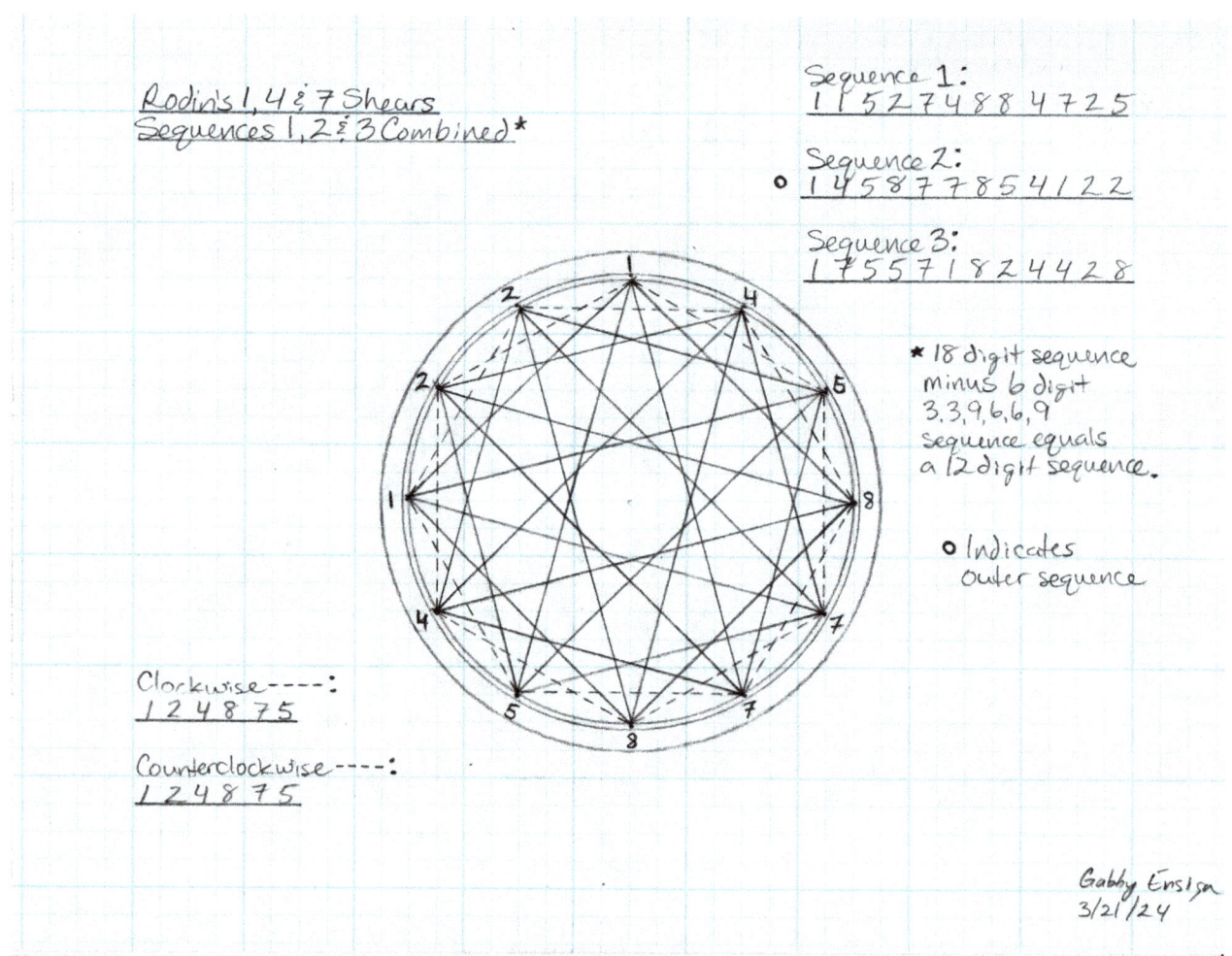

Clockwise ----:
1 2 4 8 7 5

Counterclockwise ----:
1 2 4 8 7 5

Gabby Ensign
3/21/24

Rodin's 1, 4 & 7 Shears
Sequences 1, 2 & 3 Combined*

Sequence 1:
1 1 5 2 7 4 8 8 4 7 2 5

Sequence 2:
1 4 5 8 7 7 8 5 4 1 2 2

Sequence 3:
1 7 5 5 7 1 8 2 4 4 2 8

*18 digit sequence minus 6 digit 3,3,9,6,6,9 sequence equals a 12 digit sequence.

○ Indicates outer sequence

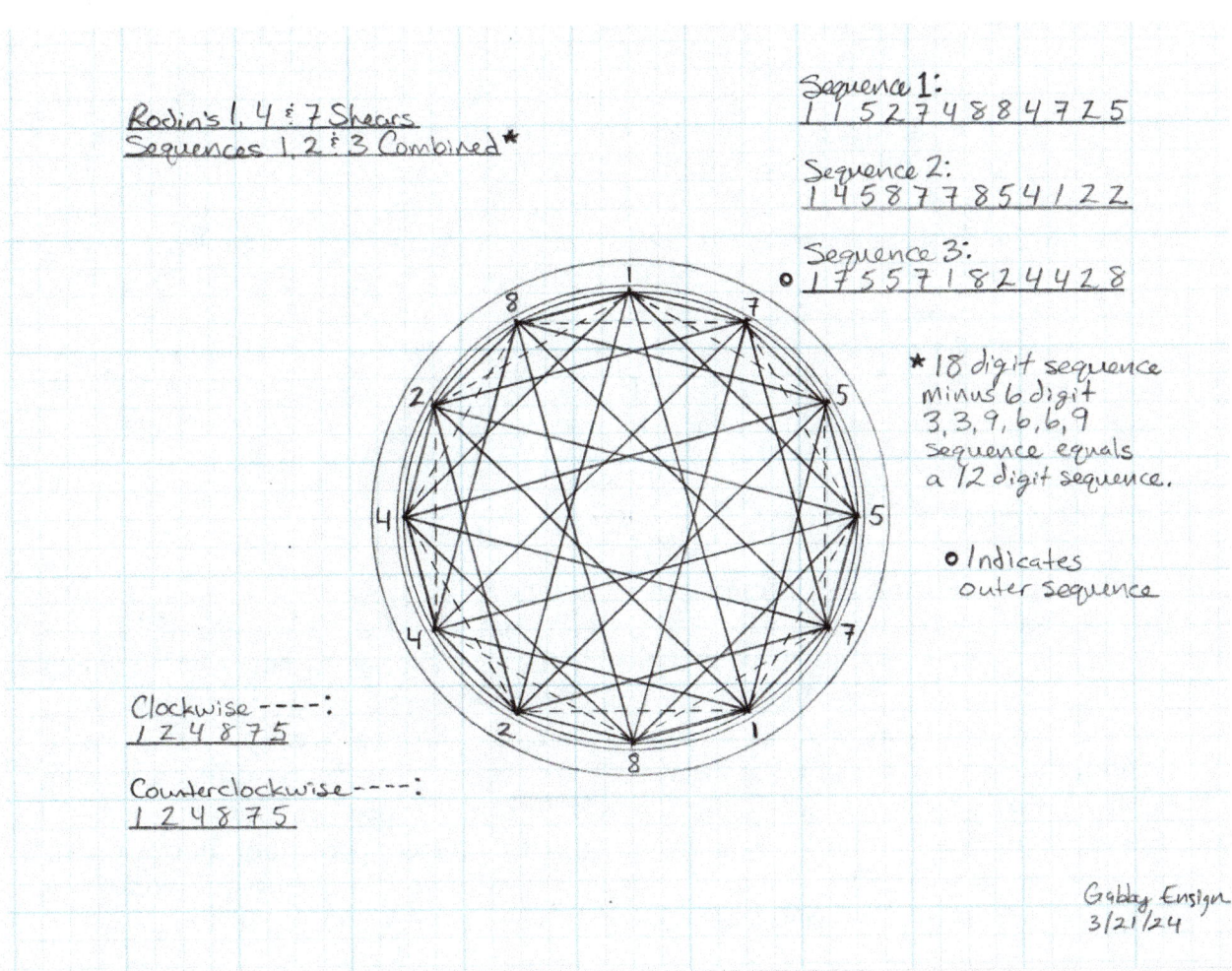

Clockwise ----:
1 2 4 8 7 5

Counterclockwise ----:
1 2 4 8 7 5

Gabby Ensign
3/21/24

The outer circle contains a 12-digit sequence and is represented by a small black circle showing which of the three sequences is making up the circle. No matter which sequence makes up the circle, they all use the exact same geometric patterns to create the same thing, but combined only. With only one and without the others, you don't get full symmetry. It is as if all three are actually part of one whole and need each other for their function to work.

Interestingly, these sequences are missing certain lines that the Diamond of Fifths has in its geometry. I have added them with dotted lines and as you can see, they are two counter-rotating 1 2 4 8 7 5 circuits. I can't say why for sure, but it is still speaking the language of vortex based mathematics, so I hope I can give you an explanation for it someday. For now, I am confident that its full geometry as seen in the Diamond of Fifths in Phi is representative of the whole geometry. If that's the case, these missing lines show me these 3 sequences do need something else working in tandem in some multidimensional fashion. Even though they all work as a whole within the three, what we see in Phi's sequence seems to be doing something to complete the geometry. It could be as simple as creating movement, which is what Phi does. It's an infinite inward and outward expansion and contraction. Imagery is symbolic of life, a picture is worth a thousand words, or when Art imitates Life, its mimesis.

The Language of Creation

These diagrams do have something to say, and if we were wise, we would try to listen. If the universe speaks, we should ask, "What would you have to say?" I have been trying to listen, and I believe it is saying, "Learn to speak my language." I would interpret this in two ways. One, learn to understand the mind of God, and two, learn to understand the heart of God. One is wisdom, the other is love. Spirit contains all of both aspects, but in a physical existence here, we have mind, body and spirit. The Trinity allows Spirit's extension in this world. Our spirit's experience through the body alters our spirit's perception here, but our spirit can correct perception if our minds and body are in sync with it. The Trinity is needed here in the physical world, but without the illusion of our physical bodies and physical minds, we are all one.

Chapter 4

The Musical Matrix of Creation

All is Frequency

Science has unequivocally determined that everything in the universe is in a state of movement and vibration. Not only is our planet rotating around the sun, the solar system rotates through the galaxy. But when we look even closer, atoms are vibrating, spinning, moving, and are being held together by pure energy. Einstein proved this by "splitting the atom," and we know what happens when we do that.

We can measure wave patterns, produce heat with friction, and we can see x-rays to look at our own bodies from the inside out. These waveforms are frequencies that have absolute characteristics and can be measured as hertz, the time it takes a current to change direction per second.

The right frequency at a high enough amplitude can shatter a crystal wine glass with the power of one's own voice. We can even turn a liquid to a solid by changing the temperature in the room and watch water freeze. Frequency and vibration are fundamental to the way the universe works.

Cymatics

We also know that form and matter take on geometric properties at all scales and sizes, big and small. Frequency and geometry together create form.

One of the best examples of this is seen in cymatics. This is the study of vibrational phenomenon when applying frequencies on a flat plate to see displacement patterns created based on the frequency of vibration in hertz. In other words, you can get perfect geometric patterns by pouring salt on or water on a metal plate, and vibrating it to certain tones.[8]

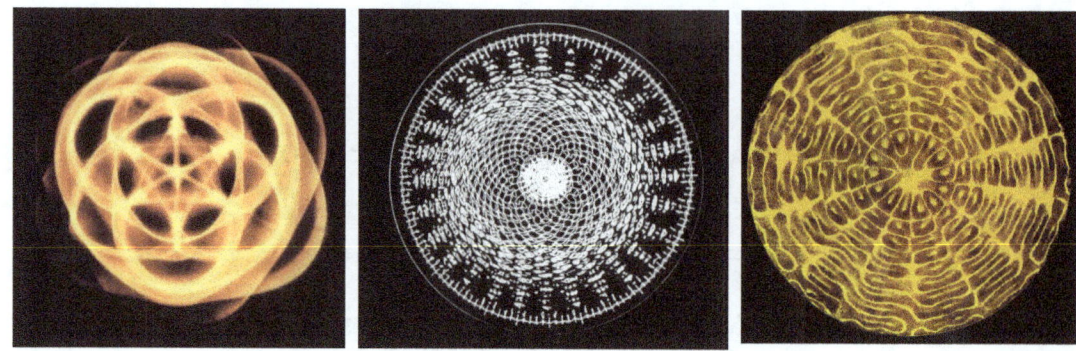

Image 25 – 27: Cymatics by Hans Jenny

Each pattern is associated with a specific tone of frequency and will create the same geometry every time that tone is used. The first time I saw this in real time it absolutely blew me away. I remember thinking, "That's it, right there, frequency creates all the geometry." I already knew it, but seeing it like I did with cymatics made it that much more real to me.

How can a frequency contain so much information regarding such intricate patterns?

The Geometry of Planets

I believe that within the frequency itself, there is a type of binary code if you will. It is the mathematical law of ratios. The numbers speak in terms of geometry, therefore numerical resonance is also a factor to consider. Another example? This next one is out of this world, literally. It is on Saturn and takes up the entire north polar region in a massive hexagon within the atmospheric activity.[9]

Image 28: Hexagon on Saturn by NASA

We also know that if you put a tetrahedron in a sphere with one point at a pole, the three other points will touch the surface of the sphere at 19.5° above or below the equator, depending if it is aligned to north or south.

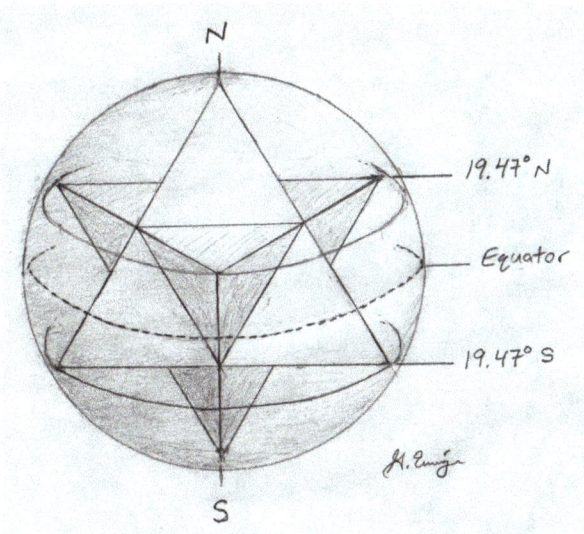

Image 29: Tetrahedron in a Sphere

This is nothing other than a geometric fact to be measured and calculated until we find some interesting observations at or very near 19.5° above or below the equator of nearly every planet in our solar system, including our sun.[10] So, let's start with the sun. The sun spots on the sun rarely go above or below 19.5° north or south of the equator.

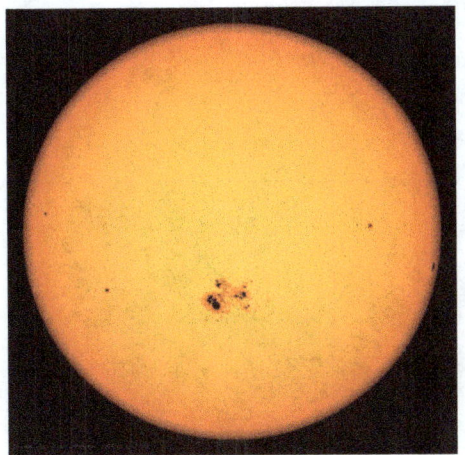

Image 30: Sunspots

On Venus, the greatest volcanic activity on the planet in the alpha and beta regions are near 19.5°.

On earth, it is Hawaii with Po'haku Hanalei being the largest shield volcano on earth at nearly 19.5° north of the equator.

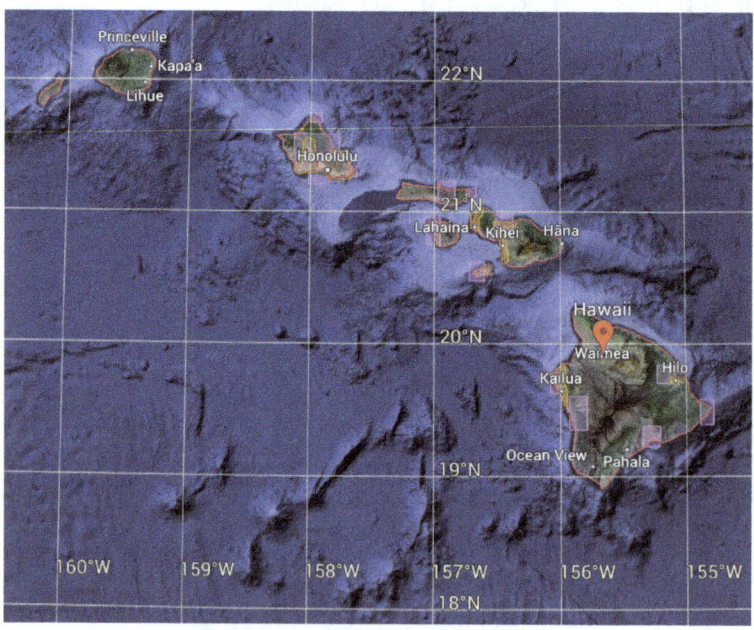

Image 31: Hawaii

Mars has Olympus Mons, which is the largest volcano in the solar system, and sits near 19.5° north of its equator.

Image 32: Mars Olympus Mons by NASA

The Great Red Spot on Jupiter is fairly stationary at around 19.5° south of its equator. Even though it is a gas planet, it is always in the same spot.

Image 33: Jupiter by NASA

On Neptune, there is a massive vortex of clouds similar to that on Jupiter, and it holds its position close to 19.5° south of the equator. It is called the Wizard's Eye.

Image 34: Neptune by NASA

All of this is not a coincidence because geometry tells us that upwellings of vortex energy manifest on the planets at a key node of a tetrahedron in a sphere. The planets resonate with this geometry. In fact, there are many other geometries. This is just one phenomenon of one form. The icosahedron can be seen on earth's grid through vortex activity on planes and ships. A researcher by the name of Ivan T. Sanderson found that if all recorded disappearances of planes and ships were plotted on a globe they would cluster around 12 equally distant points on the planet.[11] The Bermuda triangle is one of them, and Japan's Dragon's Triangle is another. The Japanese government even forbids travel through this area. Altogether they form an icosahedron.

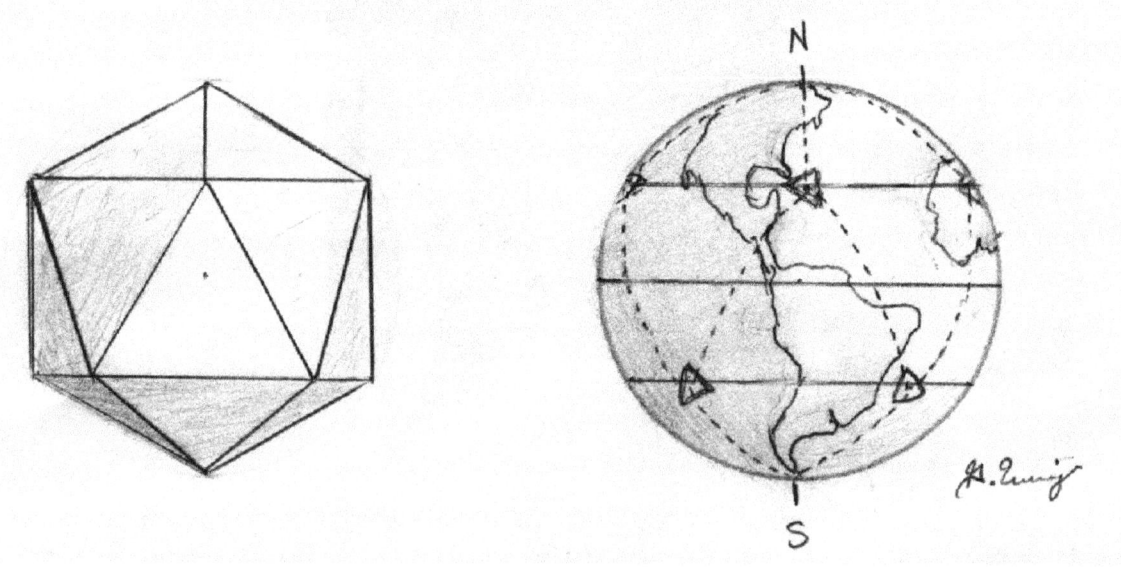

Image 35: Icosahedron as the 12 Devil's Graveyards

The Harmonic Numbers and the Circle of Fifths

There are many more examples, but this shows you the inherent role that geometry plays in creation, and frequency is the key behind it. Being a musician, I automatically equated this to tone, notes, rhythm, and music, because music is also structured mathematically and even carries properties of geometry.

So, let's talk frequency, shall we? Jamie Buturff is a vortex based mathematics researcher and has made several contributions to the field. One of which is a chart containing all of nature's most harmonic numbers. He created the following chart based on the multiplication and division of the numbers two and three, and what you see is my re-creation of it.

12 Musical Notes in Resonant Hz Frequency:
Quantifying as the Circle of Fifths

Current Hz:	261.626	391.995	293.665	440.000	329.628	493.883	369.994	277.183	415.305	311.127	466.164	349.228
Actual Hz:	256	384	288	432	324	486	364.5	273.375	410.0625	307.546875	461.2203	345.9052
Circle of Fifths	C	G	D	A	E	B	F#/Gb	C#/Db	G#/Ab	D#/Eb	A#/Bb	F
	.000977	.00243	.008789	.029367	.079102	.233305	.711914	2.135742	6.407227	19.221678	57.665041	172.9951
	.001953	.005859	.017578	.052734	.158203	.474609	1.423828	4.271448	12.814453	38.443355	115.3301	345.9902
	.003906	.011719	.035156	.105468	.316406	.949218	2.847656	8.542296	25.629	76.8867	230.6602	691.9805
	.007813	.023437	.070312	.210937	.632812	1.898437	5.695312	17.085914	51.25782	153.7734	461.3263	1,383.96
	.015625	.046875	.140625	.421875	1.265625	3.796875	11.390b	34.17187	102.5156	307.5469	922.640	2,767.92
	.03125	.093075	.281125	.84375	2.53125	7.59375	22.78125	68.3437	205.013	615.0938	1,845.28	5,535.84
	.0625	.1875	.5625	1.6875	5.0625	15.1875	45.5625	136.6875	410.0625	1,230.1875	3,690.56	11,071.68
	.125	.375	1.125	3.375	10.125	30.375	91.125	273.375	820.125	2,460.375	7,381.125	22,143.37
	.25	.75	2.25	6.75	20.25	60.75	182.25	546.75	1,640.25	4,920.75	14,762.25	44,286.75
	.5	1.5	4.5	13.5	40.5	121.5	364.5	1,093.5	3,280.5	9,848.5	29,524.5	88,573.5
	1	3	9	27	81	243	729	2,187	6,561	19,683	59,049	177,147
	2	9	18	54	162	486	1,458	4,374	13,122	39,366	118,098	354,254
	4	12	36	108	324	972	2,916	8,748	26,244	78,752	236,196	708,588
	8	24	72	216	648	1,944	5,832	17,496	52,488	157,464	472,392	1,417,176
	16	48	144	432	1,296	3,888	11,664	34,992	104,976	314,928	944,784	2,834,352
	32	96	288	864	2,592	7,776	23,328	69,984	209,952	629,856	1,889,656	5,668,704
	64	192	576	1,728	5,184	15,552	46,656	139,968	419,904	1,259,712	3,779,136	11,337,408
	128	384	1,152	3,456	10,368	31,104	93,312	279,936	839,808	2,519,424	7,558,272	22,674,816
	256	768	2,304	6,912	20,736	62,208	186,624	559,872	1,679,616	5,038,848	15,116,544	45,349,632
	512	1,536	4,608	13,824	41,472	124,416	373,248	1,119,744	3,359,232	10,077,696	30,233,088	90,699,264
	1,024	3,072	9,216	27,648	82,944	248,832	746,496					
	2,048	6,144	18,432	55,296	165,888	497,664	1,492,992					
	4,096	12,288	36,864	110,592	331,776	995,328						
	8,192	24,576	73,728	221,184	663,552	1,990,656						

÷2
×2
×3

Middle C

Starting with the number 1, it is multiplied by 2 going down column 1, and multiplied by 3 going across the row to the right. Above the number 1, everything is divided by 2. So, the numbers are doubled or halved going up or down, and tripled going right. Also, if you take the root sum of the numbers in the first column, they repeat the 1 2 4 8 7 5 doubling circuit, while the second column repeats 3 and 6. Everything right of that has a root sum of 9.

The first time I saw this chart I started seeing a lot of the same numbers I have been seeing in sacred geometry over the years, like 72, 108, 144 and 432, to name a few. That was interesting, but something else I saw was what really got my attention. The darkened areas highlight numbers that are strikingly close to the hertz value of the 12 musical notes, with our A note being 440 hertz. The top of the chart shows the current hertz assignments and the chart's hertz assignments to see how close they are side-by-side. Also, in each column are all of the octaves of each of the 12 notes and they naturally arrange into the circle of fifths in music theory. The circle of fifths arranges each note in fifths, or fourths, depending which direction you go. And again, the reason for this arrangement is because the notes closest to each other in correspondence sound pleasing to the ear when played together or in succession. The further away from each other on the chart, the more dissonant each note sounds in relation to each other. So simply put, the circle of fifths in music relates notes that are harmonically stable and consonant with each other and fall within a 3:2 ratio.

Tuning Frequencies

This chart chose the ideal frequencies for all notes within a scale when talking in terms of the way the universe structures music and sound. At a glance it may seem that we found a solution to equal temperament tuning, which is slightly out of tune, mathematically speaking, and very out of tune, harmonically speaking, relative to the hertz value of this harmonic chart of numbers. Basically, we are trying to get an 8 note octave (12 note harmonic series) to be consistent with itself in a system containing thirds and fifths. Without getting complex, it is like squaring the circle in math or trying to fit a square peg into a round hole.

Pythagoras tried to fix this problem with his 3:2 ratio tuning with something called the Pythagorean Comma, which closed the gap to the octave note, successfully doubling the octave. If an octave is a ratio of 2:1 between notes, i.e., C256 to C512, and it is impossible that a 3:2 ratio of fifths will perfectly be divisible by a 2:1 ratio, the gap is closed by a minor adjustment in tone. Our solution for the Pythagorean Comma is Equal Temperament tuning of 12 equally spaced notes, although slightly off from Just Temperament tuning. To achieve Absolute perfect tuning, it would take a piano with multiple levels of undertone keys of notes within an octave of at least 30 notes. The number 12 is simpler and more practical, so we divide by 12, which is imperfect, but close.

Musically and mathematically, we have had a problem with tuning our instruments. All systems derived have either been consonant or dissonant, and consistent or inconsistent, but never consonant and consistent. Always it is some other combination thereof. It is even said that it is not possible to create a tuning system to sound both consonant and consistent at all times. The compromise is to be consistent while playing all but the root keynote out of tune to sound pleasing or consonant. Just Temperment tuning only works with sequential notes, but not chords.

You can learn more about this problem within music and math by searching Equal Temperament, Just Temperament, and Pythagorean tuning. It can get quite confusing, as you can tell, but it isn't too difficult to understand if you take the time to learn it.

A mathematician and researcher of vortex based mathematics, Robert Edward Grant, devised a tuning system, of which I have yet to fully understand, that although it bases itself in A432, it does not match the ideal seen in the chart of most harmonic numbers. But that's OK because they are so close, needing just minor adjustments and they resonate better to our ears. Just Intonation, like what we see in the chart, is the precise, mathematical, ratio ideal; the masculine aspect of nature seen as a line. The feminine aspect as a curve doesn't operate within straight lines and twists or turns, taking a curved path.

Grant calls his tuning Precise Temperament and devised it by calculating to its closest degree the square of a circle.[12] He explains it much better than I can so I recommend watching his video on YouTube, *"The Case for Precise Temperament Tuning in 432.081 Hz Instead of Equal."* Grant isn't just another voice in the crowd of theories, he is among the leading candidates offering a unified field model that answers many problems in physics and science. In fact, between Marko Rodin, Robert E. Grant, Nassim Haramein, and Malcolm Bendall, we have a unified field model, because each of them fill in some of those missing pieces to the puzzle.

The Musical Notes of Phi?

All that being said, the musical notes still arrange themselves into 12 primary tones of the circle of fifths sequence, even though there are variable undertones to make up all harmonic tones in music. It makes sense if all things are not always how we see or hear them in these physical domains we occupy. Music could also express itself in more than three dimensions.

The geometry of 12 notes does fit within our double hexagon of 6, our half of 24 with the day in hours, and within the chart of most harmonic numbers. The geometry of 12 also fits within Phi which is half of the 24-digit sequence in its entirety, working as two circuits of 12 digits. This creates the toroidal numerical relationships within Phi and vortex based mathematics.

All in the universe is created from a toroid and therefore, frequency, vibration and tone are also inherent within this form. The three sequences of the Rodin 147 Shears and Gator's Donut models the toroidal arrangement of interlocking sequences, when reduced from 18 digits to 12 by removing all of the 3's, 6's and 9's. I discovered that with 12 it creates the geometry of the circle of fifths in Phi. A toroid itself functions as an expression of Phi. Everything pleasing in nature contains Phi. Music is no different and neither are its notes. One could say love is Phi.

This next diagram of notes and hertz based on the numbers of Phi is only my theory at this point, and it is probably the only part of this book that primarily relies on inherent logic in their assignments, other than empirical data. Either way, the 12-digit diamond of fifths pattern from Phi that I discovered and the circle of fifths that Pythagoras and others before him found both carry the same geometric sequences. The question I ask myself is, "Is there a relevant connection?" I believe there is, but this theory regarding musical notes and their numerical assignment based on Phi is yet to be validated, so take it as it is and do not use it if it does not conform to your understanding. This would be like assigning notes to the feminine curve of the toroid, instead of the cuboctahedron or vector equilibrium, which is the line and angled male aspect. I find the relationships interesting.

The ratios derived within this next diagram were calculated from the notes assigned in the chart of most harmonic numbers. The point in time when I learned these ratios of Pythagorean tuning came a while after I attempted to incorporate a numerical assignment to each of the notes using the sequences and geometry of Phi. Since Phi created the same geometry as the way harmonic music was arranged, both in 12, there might be a correlation between the two. It turns out it is similar to how Pythagoras came up with his system of tuning, but without the Pythagorean Comma. Well almost, my sequence of ratios mirrors itself down the middle when moving into the sharps/flats, whereas his relied on a more sequential calculation of 3 over 2; multiplying or dividing by 1.5, and by multiplying by .5 to go down an octave. The difference is that mine isn't looking for Consonant, only Just Tuning, because the feminine curve of Phi does what it needs to do with the Just Tuning ideal in order to make it a balanced system within the whole. It is titled, *"Musical Notes and Their Hertz Relationship to Each Other,"* and is finding exact mathematical ratios. It is not trying to fit within a system of tuning instruments, merely showing geometry in its delivery of the harmonics of 12 within balanced hertz relationships, and seen as notes aligning themselves in the circle of fifths pattern.

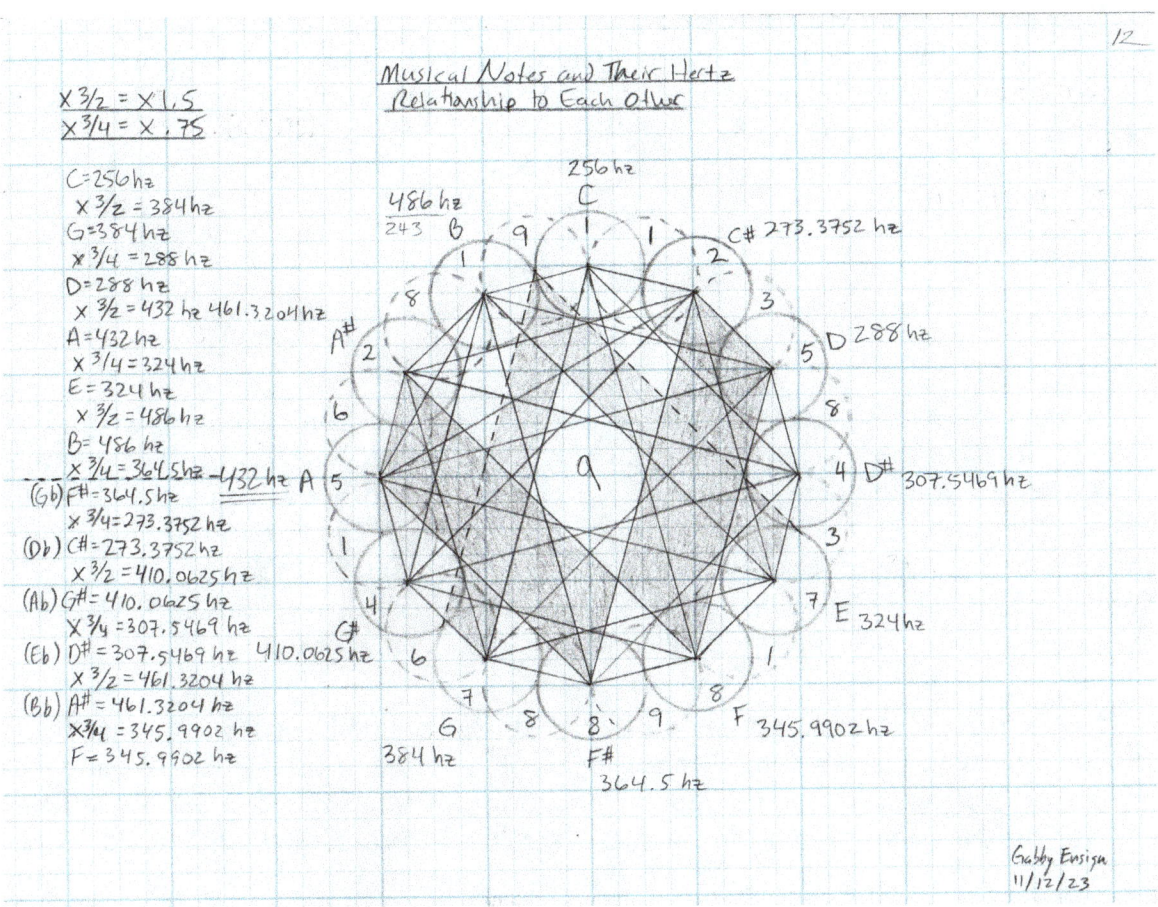

The relationships between each of the notes are laid out on the left side of the diagram. You can see my alternating multiplication sequence of multiplying by 3/2 and 3/4, then reversing from 3/4 to 3/2 in the middle as the notes move into sharps. It is the point of reversal in some sense just as 8 is the middle number reversing the sequence in the opposite direction. Everything in vortex based mathematics is polarized it seems, so hertz relationships show that attribute as well. Interestingly enough, it is the exact same way the 12-digit Phi sequence from the Diamond of Fifths has a half point reversal if assigning the first digit in the sequence to the note C. Ha, ha … C my logic? Well maybe it is not logic I see at this point, but going where the numbers tell me to go, it seems there is a correlation. Whether if it is relevant we shall see, it fits within Just Tuning or ideal in ratio terms and geometrically within its relationship to the octaves and the circle of fifths. Let's investigate further.

The Family Number Groups

Within vortex based mathematics, you should now see a little more evidence of the 147 and 852 family number groups. The Rodin 147 Shears grid maps and Gator's Donut grid map have them showing up nicely when you stack the sequences next to each other as such (minus the 3 3 9 6 6 9 circuit).

1 8 2 4 4 2 8 1 7 5 5 7

1 5 2 7 4 8 8 4 7 2 5 1

1 2 2 1 4 5 8 7 7 8 5 4

In between each set of 3 numbers, which are the same going from top to bottom, repeat these numbers sequences: 852, 471, 285, 147, 528 and 714. The number sequences that are the same are 1, 2, 4, 8, 7 and 5 in that order, which is the doubling circuit. Take another look at Gator's Donut on the next page in a toroid grid. Down the center is 9, then going clockwise from center is 528, then 528, then 6, then 147, 714, 6, 852, 285, 9, 714, 147, 3, 285, 852, 3, 471, 471, then starting over back to 9. These are the same family number groups divided by a ribbon of 3, 9, or 6.

Nowhere in this arrangement do we see the two family number groups being grouped between two 9's. Only between 9 and 6, and 9 and 3, whereas 147 and 714 are shown between two 6's, and 285 and 852 are shown between two 3's. When I saw this for the first time the phrase ringing in my head from something I heard Marko Rodin say was, "…3 governs 8, 2 and 5, whereas 6 governs 1, 4 and 7." I was pleasantly surprised when I saw an indication of this within Gator's Donut because it helped validate the inherent vortex based mathematics in it.

How does this relate to the musical note's hertz number assignments? Just wait, I'm getting there.

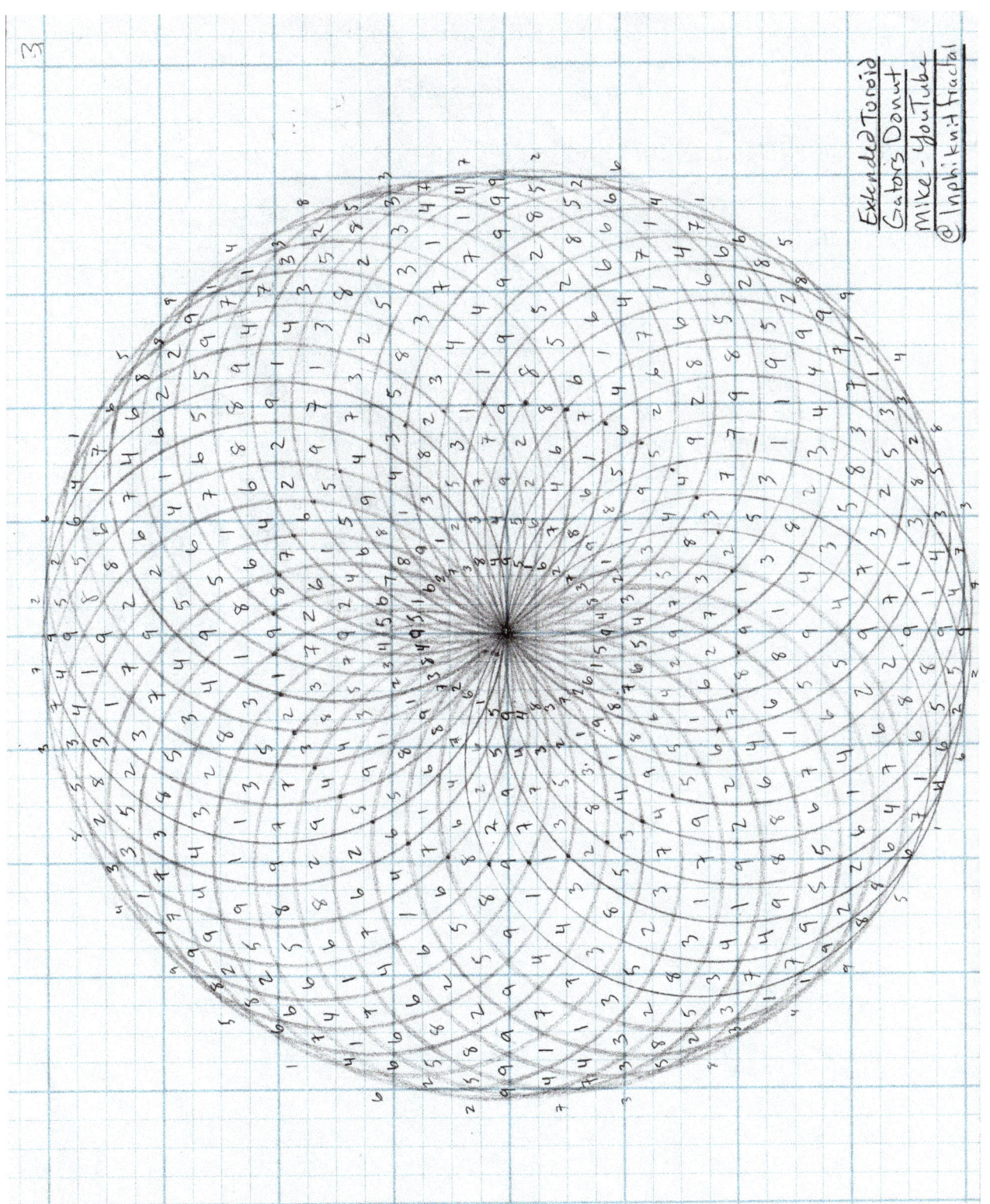

More examples of the 147 and 852 family number groups will become very apparent throughout the rest of this book, but the idea that 3 governs 852 while 6 governs 147 is an interesting synopsis that leads to further intrigue.

Vortex Based Mathematics and the Flower of Life

Since life expresses itself through geometry, what better place to look then at the first stages of life when a single-celled organism doubles for the third time to become a total of 8 cells. This is the first time a lifeform has completed all of the single digits 1 through 9. Well actually, 9 outward to the rest of the 8 because 9 is the center, 9 is the source of life, and 9 is unity before the split, but I digress, you get the idea. If we number the cells with 3 and 6 as the polar axis and 1 2 4 8 7 5 going across its central axis, the doubling circuit along with the family number groups and the 3 and 6 governorship become apparent in their relationships. See the following diagram, or watch a three-dimensional graphic rendering of this on YouTube from @NFGC Tarot, in a video titled, *"Vortex Based Math & The Flower of Life Discovered & Copyrighted by Bill Wandel"* which I highly recommend.[13] He does a much better job showing you its function in a video than I can on paper. That's OK though, because my graphic will give you enough of an idea without watching the video.

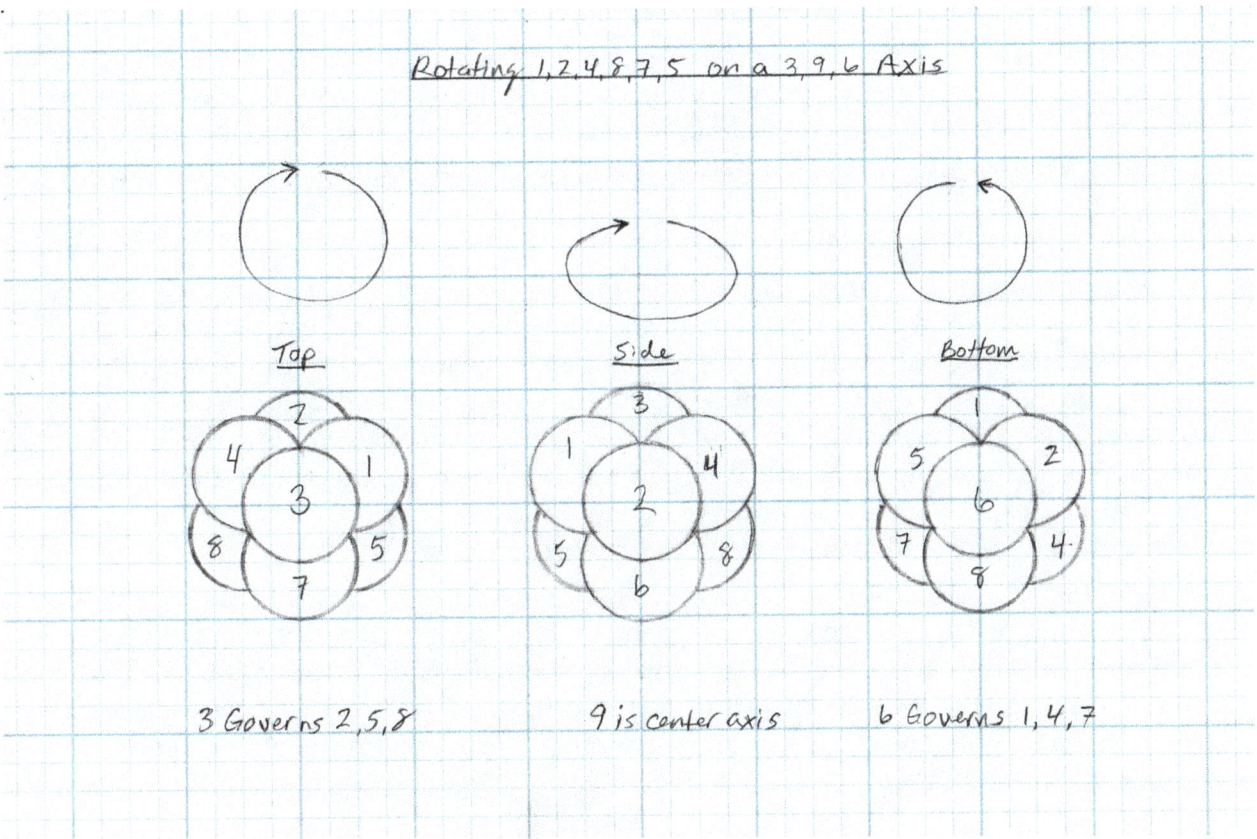

The first image looking downward has 3 on top, 1, 4 and 7 just below that, then 8, 2 and 5 on the lower half, and 6 at the bottom which you cannot see. The middle image shows the side view with 3 on the top, 6 on the bottom, 1, 4 and 7 in the back on the upper half, and 8, 2 and 5 on the lower half. The same goes for the third image except 6 is showing its reverse. The interlacing geometries that connect the center points of each cell are in a double tetrahedron or star tetrahedron arrangement.

With each point corresponding to a number, we get the numbers 3, 8, 2 and 5 in one tetrahedron, and 6, 1, 4 and 7 in the other. The relationship with the math tells us this is a perfect arrangement for the numbers 1 through 8. Looking from the top, 1 + 4 + 7 = 12 = 3, looking from the bottom, 8 + 2 + 5 = 15 = 6, and looking from the sides, 1 + 4 + 6 = 11 = 2, 8 + 5 + 3 = 16 = 7 (in the back), or 3 + 2 + 5 = 10 = 1 looking around the side. Any relationship between nested numbers of three result in a sum of the number at its center. Fabulous, seeing as it is being comprised of two separate forms with two different sets of numbers. There is that polarity and reciprocation thing again.

Hertz Relationships

To establish the apparent relationships between 3 and 825 with 6 and 147, it was important to show you these diagrams first before moving to the next piece and comprising a musical note assignment to the 12-digit half sequence of Phi containing the diamond of fifth's.

The next diagram arranges the family number groups with their governing number of 3 and 6, while showing the mathematical relationship between the two notes assigned to each number, along with their hertz value and polarity at any given time. When looking at the previous chart showing the notes with Phi, C and B correspond to a 1, A♭ and E♭ to a 4, E and G to a 7 and so on. Then each arrow corresponds to the balancing or flipping nature of each note moving to its closest neighbor within the circle of fifths. It also shows the corresponding ratio in hertz of 3/2 and 4/3 (the mathematical inverse of 3/4 as seen in Pythagorean tuning).

Beginning with C in the circle of fifths with 1, we give it a polarity, in this case negative (-), and go to G which is positive (+) and has a hertz ratio of 3/2. D is now negative going to the A being positive in another 3/2 ratio on its way to E being negative and so on. Every time it goes between a 3 or 6 governorship, 9 is quietly the center of the circuit.

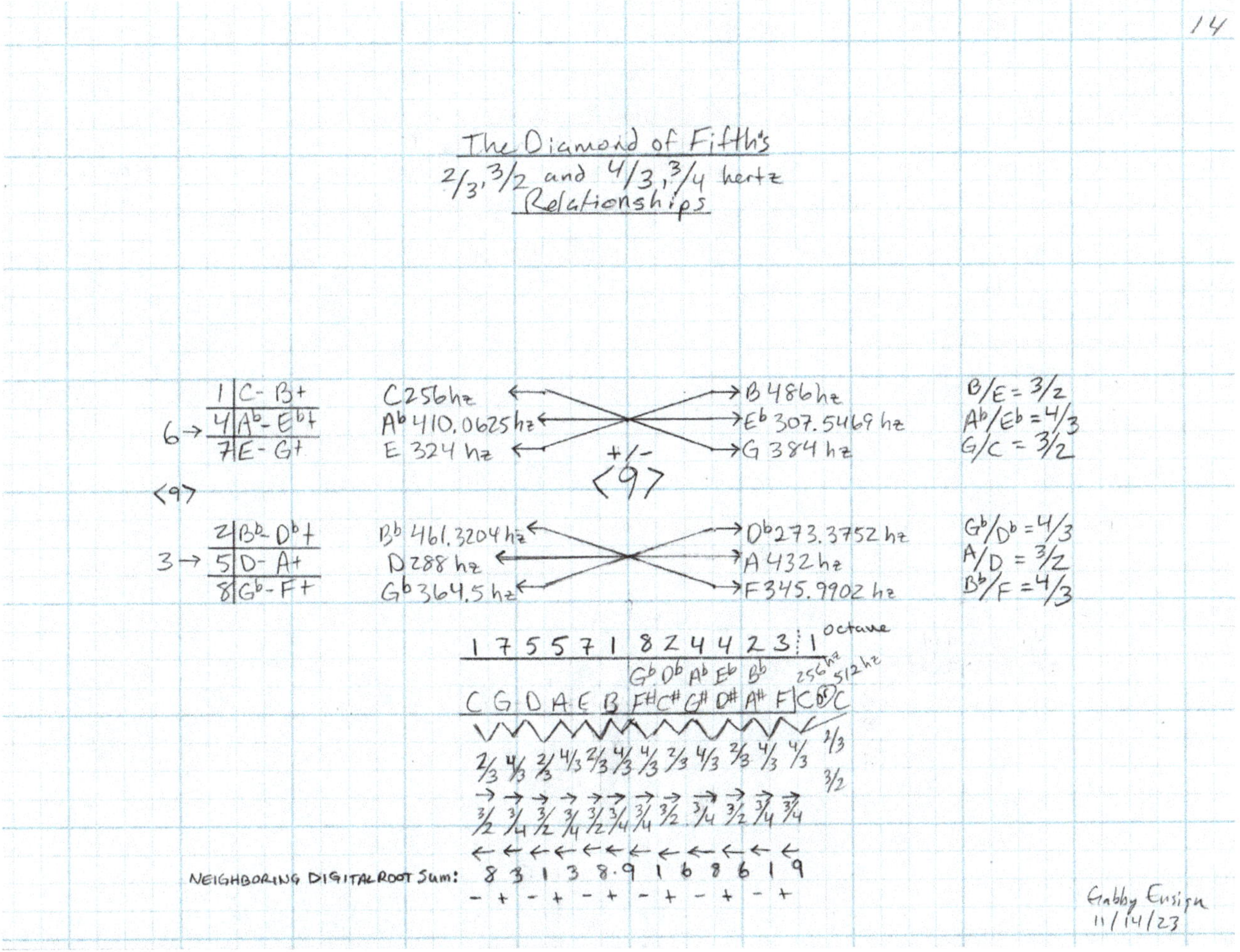

The Diamond of Fifths
2/3, 3/2 and 4/3, 3/4 hertz Relationships

[Handwritten diagram showing:]

6 → | 1 C- B+ | C 256 hz ←——→ B 486 hz | B/E = 3/2
    | 4 Ab- Eb+ | Ab 410.0625 hz ←——→ Eb 307.5469 hz | Ab/Eb = 4/3
    | 7 E- G+ | E 324 hz ←——→ G 384 hz | G/C = 3/2

⟨9⟩   +/-   ⟨9⟩

3 → | 2 Bb- Db+ | Bb 461.3204 hz ←——→ Db 273.3752 hz | Gb/Db = 4/3
    | 5 D- A+ | D 288 hz ←——→ A 432 hz | A/D = 3/2
    | 8 Gb- F+ | Gb 364.5 hz ←——→ F 345.9902 hz | Bb/F = 4/3

1 7 5 5 7 1 8 2 4 4 2 3 1 octave
        Gb Db Ab Eb Bb 256hz 512hz
C G D A E B F#C# G# D# A# F|C|C
2/3 4/3 2/3 4/3 3/2 4/3 4/3 3/3 4/3 2/3 4/3 7/3
3/2 3/4 3/2 3/4 3/2 3/4 3/2 3/4 3/2 3/4 3/2 3/4

NEIGHBORING DIGITAL ROOT SUM: 8 3 1 3 8·9 1 6 8 6 1 9
                             - + - + - + - + - + - +

Gabby Ensign
11/14/23

The same flip-flopping is being seen here, not only between the note polarities and their assignments like C is to G, and D is to A, but as well between the back-and-forth reciprocation between 3 and 6, moving to the 9 in between to maintain movement between the polarities as seen in the Rodin symbol diagram. If we look carefully, another sequence we saw earlier is also being seen. The circle of fifths goes through two notes within a governorship before moving to 9 onto the opposing governorship. If C being 1 is in 6, then G being 7 in 6, the number 6 is seen twice before moving through 9 to return to 3. And 3 does the same thing. This creates a 6 6 9 3 3 9 circuit.

With all this being said, these assignments of notes to numbers being derived from a very logical and intuitive placement of the circle of fifths geometry overlaid with the sequences in Phi, which I now call the Diamond of Fifths and The Trinity of Creation, is theoretical and needs further research to be validated in this exact form. Nonetheless, the frequencies are there within the fabric. This is just my best attempt at understanding the 4/3 and 3/2 hertz relationship that music might have with Phi.

Chapter 5

The Music of the Spheres

Pythagoras

Pythagoras lived from about 570 BC to about 490 BC. He is sometimes referred to as the first philosopher, although there were ancient forms of philosophy before him. He was the earliest, most prominent philosopher in Western history. Others that followed found their roots in Pythagorean philosophy, like Plato, around 400 BC, with the five Platonic solids, and Hermes around 200 BC, with the Hermetic Texts.

Pythagorean philosophy was centered on the idea that numbers, mathematics and geometry made up the fundamental framework of reality. This idea was the basis for the music of the spheres in the Pythagorean school of thought. It sounds a little bit like the whole field of sacred geometry. He saw the perfect harmonic relationships with mathematics and geometry as a function of music, which in turn serves as a function in all creation, like the universe and the planets in our solar system. Pythagoras was absolutely right in that music today is clearly understood as a system of perfect mathematics and harmonic ratios. This is why Pythagoras and his followers looked at the planets and associated them with music, hence the term, music of the spheres.

The Song of Life

The harmony of mathematics is inherent in the harmony of music, and this is why a perfect fifth in music sounds so appealing to the ear. It just works, like everything else in nature and the universe. When harmony is present in numbers and geometry, it must be present in music too.

Since everything in the universe is vibrating, and is inherent with some kind of frequency, and frequency is tone, and tone is the hertz relationship to sound and musical notes, and notes in harmonic frequencies produce geometry as in cymatics, then we can say that everything in the universe is singing a type of song. It is the utmost song of life, because this whole universe is a hologram of focused energy and nothing created here even really exists at all. Call it an echo of the Voice of Creation, all within the Mind of Creation. The only "real" thing is you, which is the light coming through to occupy the physical experience. That life is first of the creator, who in turn gave life to all other things. That life is also the animals and insects of the Earth. That life is the grass and the flowers, the trees and its moss. That life is spirit, consciousness, and is the essence of love itself. The song of mathematics and geometry was written by the creator, for us to play a part in the symphony of life here on earth. But the song of life exists beyond the song of the universe. It is the song of life that created the song of the universe, and is the breath behind it.

The Number 12 Throughout History

Let us start with a little history lesson.

I have already mentioned the first prominent number of three and a few examples of religious triads in Chapter 3. There are many more; I just gave you a few examples. We also have a similar thing happening with the number 12. Without even starting on the list below, I'm sure you already have a few of them in mind as you are reading these words. Here is what I found with the number 12.

12 months in a year
12 inches in a foot
12 notes in the musical scale
12 eggs in a dozen
12 houses of the zodiac
12 days of Christmas
12 apostles of Jesus
12 sons of Ishmael
12 sons of Jacob
12 tribes of Israel
12 sons of Odin
12 labors of Hercules
12 Olympians of the Pantheon
12 Lingas of Shiva

I am sure I missed some, but as you can see, the number 12 is extensively used throughout history and is even a part of our system of measurement in the United States and other countries. Unfortunately, the metric system is the dominant one since French mathematicians created it around 1790 and, it was made into law by the French government in 1795.[14] Prior to that, the ancient structures and monuments of the world show vast amounts of evidence that the 12 inch foot was being used.

The 12 houses of the zodiac have been around since the Persian era circa 400 B.C.[15] The rest is history, shall we say? The number 12 has been around for millennia in cultures around the world.

The Number 72 Throughout History

The next number of prominence I find is 72. Mathematically it is the number 36×2 and 36 is the basis for the circle of 360°. Also, the five interior angles of a pentagram are 36° each, and the five outer angles of a pentagram are 72°. It is also half of 144, which is a number that is found all throughout the Bible and in other religious texts. Here is the list for 72 I have found, and again I'm pretty sure there are more to add to this list. These are just the most prominent ones I found.

72 names for God in the Kabbala

72 immortals in Taoism

72 disciples of Confucius

72 martyred companions of Zarathustra

72 spoken languages at the tower of Babel

72 conspirators sent by Set to kill Osiris

72 major temples at Angar Wat

72 stupas at the world's largest Buddhist Temple at Borobudur

72° slope angle of Jacob's ladder in the book of Genesis

It is easy to see why anything to do with harmony of life, in this case, numbers, is reflected in religious and spiritual traditions. Oh, but this isn't all there is. Don't worry, this all ties into the music of the spheres. I'm just showing you what is here on earth first before we jump into space. Just stay grounded here with me for a few more minutes.

The Number 108 Throughout History

The pentagram is made up of the five points of a pentagon, and the interior angles of a pentagon are 108°. Let me remind you that history might scare you into thinking anything having to do with the pentagram as being evil because of the occult uses of it. Well, the Nazis turned the swastika backwards and it became known as a symbol of evil, when it had been used for centuries as a symbol for well-being, fertility and healing in India and Asia (and is also a universal symbol used by Ancient Greeks, Celts, Anglo Saxons, Tibetans, Hindus and by indigenous people in The Americas).[16] The inherent geometry that reflected all aspects of life and creation, held the meaning of the symbols. Case in point, here is our next number, 108.

108 followers of Krishna

108 beads in Hinduism and Buddhism Mala bead necklaces

108 Om Namah Mantras in Hinduism

108 spires on the wheel of Vishnu in Hinduism

108 Upanishads in Hinduism

108 Marma pressure points in the human body in Ayurvedic medicine

The Number 432 Throughout History

And another... 432, the harmonic A note.

$432 \div \pi = 137.5°$, the golden angle of Phi.

432,000 total years of reign by the 10 antediluvian Kings by Babylonian astronomer Berosus in 300 BC called the Ziusudra.[17]

	Berossos	
	Antediluvian Kings	Years of reign
1	Aloros	36,000
2	Alaparos	10,800
3	Amelon	46,800
4	Ammenon	43,200
5	Megalaros	64,800
6	Daonos	36,000
7	Eudoraches	64,800
8	Amempsinos	36,000
9	Opartes	28,800
10	Xisuthros	64,800
[=Ziusudra]		432,000

Image 36: Berosus Chart

43,200 years in 12 sars of Enmenlu-Anna rule of Sumerian antediluvian rulers.[18]

City	Ruler	Reign in years
Eridu	Alulim	28,800
	Alalgar	36,000
Badtibira	Enmenlu-Anna	43,200
	Enmengal-Anna	28,800
	divine Dumuzi	36,000
Larak	Ensipazi-Anna	28,800
Sippar	Enmendur-Anna	21,000
Shuruppak	Ubar-Tutu	18,600
		Total: 241,200

Image 37: Sumerian Chart

4.32 billion years in the Hindu Kalpa or one day of Brahma in Hinduism.

432,000 years in a Hindu Kali Yuga and 4,320,000 years in a Chatur Yuga.[19]

	Divine Years	Solar Years
Chatur Yuga	12,000	4,320,000
Krta Yuga	4800	1,728,000
Yuga Sandhya	400	144,000
Yuga	4000	1,440,000
Yuga Sandhya	400	144,000
Treta Yuga	3600	1,296,000
Yuga Sandhya	300	108,000
Yuga	3000	1,080,000
Yuga Sandhya	300	108,000
Dwapara Yuga	2400	864,000
Yuga Sandhya	200	72,000
Yuga	2000	720,000
Yuga Sandhya	200	72,000
Kali Yuga	1200	432,000
Yuga Sandhya	100	36,000
Yuga	1000	360,000
Yuga Sandhya	100	36,000

Image 38: Hindu Yuga Chart

There was definitely something very interesting going on in the past, and whether these stories are based on facts or are just stories or myths, it is interesting that they still all share the same sets of the numbers, not just the 432, but a whole host of numbers, all of which can be found on the chart of most harmonic numbers, seen in the last chapter. Even the Mayans saw harmony and numbers in their calculations of the cycles of time, like 7,200 days in a K'atun, 144,000 days in a B'aktun, and 2,880,000 days in a Piktun.[20] It is just simply amazing.

Celestial Harmonics

We can speculate that this could have been all coincidence, but I doubt that. It could have been influenced by ancient antediluvian knowledge or given to them in visions. They could have seen harmonics in the stars and constellations. There are many possibilities, but how doesn't matter so much if we know they were all expressing what is there within the harmonics of nature and the universe, because that's where the truth lies in the end. They knew the truth. We now have the answer and it matters not where they found the numbers, although I am curious. But we may never know. I personally believe that they found these numbers in the planets and stars and incorporated them into their legends and/or were actual beliefs in their cultures. Either way they were right with the numbers. Not just because we find them in our chart of most harmonic numbers, but because we actually find them in the planets and here on earth in our measurement of time and distance. I mentioned before that the ancients used a standard foot in the construction of their sacred sites. They did so because it is the most accurate way to measure. How do I know that? Let's take a look at some more harmonic numbers.

> 36: There are 3,600 seconds in an hour. The sum of all angles in an icosahedron is 3,600°. There are 360° in a compass and a circle.

× 2 = 72

> 72: 72 days on earth is equal to one day in the great year of the 25,920 year precession of the equinox. The sum of all angles in a tetrahedron is 720°.

× 1.5 (3/2) = 108

> 108: The radius of the moon is 1,080 miles.

× 1.5 (3/2) = 144

> 144: The sum of all angles in an octahedron is 1,440°.

× 1.5 (3/2) = 216

216: The diameter of the moon is 2,160 miles. The sum of all angles in a cube is 2,160°. One age of the zodiac, which is 1/12th of the precession of the equinox is 2, 160 years. The total equatorial circumference of the earth is 21,600 nautical miles. There are 21,600 minutes in a compass.

× 2 = 432

432: The radius of the sun is 432,000 miles. The speed of light at 186,624 miles per second is 432^2.

× 1.5 (3/2) = 648

648: The sum of all degrees in a dodecahedron is 6,480°.

× 1.333... (4/3) = 864

864: The diameter of the sun is 864,000 miles. There are 864,000 seconds in a day. The diameter of Jupiter is 86,400 miles. The sum of a 720° tetrahedron, a 1,440° octahedron and a 6,480° dodecahedron equals 8,640°.

× 1.5 (3/2) = 1,296

1,296: There are 1,296,000 seconds in a compass.

× 2 = 2,592

2,592: The great year called the procession of the equinox is 25,920 years, calculated by the wobble of the earth's axis by one degree every 72 years (72 × 360 = 25,920). There are 2,592,000 seconds in the average month of 30 days. The speed of light at 671,846,400 miles per hour is $25,920^2$. It is 25,920 light years from the earth to the center of the galaxy.

× 72 = 186,624

186,624: The speed of light is 186,624 miles per second.

Can you start to see how the planets are aligned with our measurement of time, the geometry of fundamental polygons, the movement of the earth, the distance to the center of the galaxy, and even the speed of light, through the whole ratios of harmonic numbers? To make it even clearer, we can find their relationship to each other by multiplying them with other harmonic numbers. For example, if we take the diameter of the moon, which is 2,160 miles, and multiply it by 108, the inner five angles of a pentagon, we get the distance from the moon to earth at 233,280 miles. The same goes for the diameter of the earth at about 8,000 miles, which when multiplied by 108 gives us the diameter of the sun at 864,000 miles. And once more, if we take the diameter of the sun and multiply it by 108, we get the distance from earth to the sun at 93,312,000 miles. Is your jaw dropping yet? These calculations are nearly all within a 97% to 99% accuracy. Go ahead and check them for yourself and divide the two to find the differences. You'll be amazed.

Although earth's diameter doesn't show up on the chart of most harmonic numbers at 7,920 miles, we can take the sum of all angles in a 720° tetrahedron, a 1,440° octahedron, a 2,160° circle and a 3,600° icosahedron, and we get 7,920. There are more harmonic ratios in our solar system, but these are the best examples to show you for now.

The only one left I'm going to show you is related to Phi and that is the speed of light at 186,624 miles per second equals 161,803.4 nautical miles per second, which is Phi minus the decimal point of 1.618034, (186,624÷1.1508 [ratio between miles and nautical miles]) = 161,874.977 which is accurate within 99.9%!

Denying the Truth

The implications for this are incredible. Why doesn't everyone know about this? Why haven't our astronomers and scientists figured this out yet? It is so easy to see once you punch in the numbers. Maybe they do see but either haven't connected the dots because the actual numbers fall within 1% to 3% accuracy to our measurements? Or maybe it is a collective denial, and thinking in these terms is too much for our astronomers and scientists, who may tend to have a strictly scientific perspective with no room for the term "intelligent design." It doesn't matter, truth always wins in the end and they can choose to put their hands over their eyes and deny its apparentness as long as they want, it won't change the fact that the numbers and the math are there. Randall Carlson, a well known geometrist, has been showing us these relationships for the last few years. I believe we used to have some of this knowledge thousands of years ago, but have since lost it, and whatever still exists is being suppressed. Hence the dominant metric system of 10 instead of the 12 inch foot.

As you can tell, I am a proponent of the 12 inch foot. Besides the obvious reasons I have just mentioned, it is because it has been used since the Sumerian civilization in ancient Mesopotamia over 6,000 years ago. According to our history books it is because of the Sumerians we have a 12 inch foot, a day split into two halves of 24 being 12 hours each, have 60 seconds in a minute and 60 minutes in an hour.[21] I am surprised it has lasted this long, considering man's constant quest to deny the truth. But that will hopefully begin to change with books like this, and others learning the truth.

Ancient Geomathematics

A gifted mathematician named Carl P. Munck discovered a system of geomathematics in the measurement, location and design of nearly every ancient megalithic structure, including ancient mounds in North America and Europe.[22] I mentioned earlier that by looking at a site's layout we can determine its precise position on the earth through its geo-coordinates of latitude and longitude based on a prime meridian of the Great Pyramid in Egypt. It is an ancient system of math, which describes the methods the ancients used to find the volume of a sphere, the area of a sphere, the area of a 360° circle, the radian, and other math constants through geomapping. Our math is different than their method, but Carl shows us that theirs was simpler, and more elegant. Along with this, he gives plenty of examples that they used the 12 inch foot and the 5,280 foot mile. Learning of all this has changed my perspective on what the ancient civilizations knew and what they were capable of doing. If they had "outside" help is up for debate, but the facts can be seen and are inherent in the work of Carl Munck. He undeniably shows us the ancients knew of precise mathematics, precise geocoordinates, and the harmonics of numerical geometry. Well, this book is full of mind blowers and gems, so get ready for another one.

As mentioned, the ancient civilizations used a prime meridian over the Great Pyramid in Egypt. Considering that the Great Pyramid is built directly at the center of all the earth's landmass, it is easy to argue that it always was earth's original prime meridian. That means that going east to west it centers the most land in both directions and the same for north and south. The Great Pyramid of Giza is at that exact spot and is the Rosetta Stone in decoding these ancient sites. Here are some of the geographic coordinates Carl found.

> 36: The Prime Meridian of 0°/360° over the Great Pyramid. The Cuicuilco Pyramid in Mexico is a circular pyramid giving equations for circular math and it is located at 19° 18' x 1.052631" N. And 19 x 18 x 1.052631 = 360°.
>
> × 3 = 108
>
> 108: The Emerald Mound in Mississippi is a 5-sided mound with 6 surfaces in a 360° poly-oval shape which equates to 5 x 6 x 360 = 10,800 = 31° 38' 9.16" N

× 1.5 (3/2) = 162

162: Manos, a glyph mound in Nazca, Peru depicts a man-like figure with 2 hands and 9 fingers with 4 fingers on one hand and 5 on the other. It has an obvious 90° angle shoulder so 90 × 9 × 4 × 5 = 16,200 = 14° 41' 28.223" S.

× 1.333... (4/3) = 216

216: Stonehenge, when completely intact, was comprised of 60 perimeter stones circling 15 inner stones, and 360 × 60 = 21,600 = 51° 10' 42.352941" W. The 15 inner stones were used for square root functions in describing their math. In Newark, Ohio there is a replica of Stonehenge's phase one circle attached to an octagon called The Fort or The Octagon. Its latitude is 40° 2' 27.00" N, which equals 2,160. It also has a longitude of 113° 34' 56.2207" W of Giza, which equals 216,000.

× 1.333... (4/3) = 288

288: The diameter of the phase one outer circle of Stonehenge was measured by British archaeologist ATC Atkinson to be exactly 288 feet. Some are accustomed to using the metric system, yet Atkinson measured it at 288 feet instead of 87.782 meters because feet was a precise whole number measurement. Going back to Peru, in Nazca, the Spiral on the Great Rectangle is a mound made up of 2 separate arcs spiraling into 4 exiting arcs which equates to 360 × 2 × 4 = 2,880 =14° 40' 5.14" S.

× 1.125... (9/8) = 324

324: Motley Mound at Poverty Point National Monument in Louisiana, is an oval-shaped mound, which is actually an eye on a mile long face outcropping the banks of a river. The face is looking due east which equates to 90°. So, 90 × 360 is 32,400, which is 32° 39' 25.961" N. Dead center on the eye... we call that a bullseye!

× 1.333... (4/3) = 432

432: In Guatemala at Tikal's Temple One, the Mayan Temple of the Giant Jaguar, its structure encodes 48 90° corners on each of the 9 terraces. That is 48 × 90 = 432 90° corners, and 432 × 90 = 38,880 and 38,800 × π = 121,893.795 = 120° 45' 22.9183" W and 239° 14' 36.9871" E of Giza. This actually calculates to 123,758.8366 which is within 98.5%. Carl Munck calls it the peripheral pyramid, because its coordinates in either direction East or West of Giza calculates into the same number. The Mayans knew the circumference of the earth at the time and even knew how to encode it into their pyramids.

× 1.5 (3/2) = 648

648: Earlier we mentioned Manos in Nazca Peru. Its grid longitude is 106° 14' 43.67" W of Giza, which is equal to 64,800.

× 1.333… (4/3) = 864

864: At Nan, an ancient city in Mexico, there is a strange pyramid made up of several million 5 to 20 ton blocks, 12 feet to 25 feet long, which are in an elongated prism shape. They are stacked log cabin style, containing more rock than in the Great Pyramid, and placed on an island that was built just for the site's construction. The prism stones are arranged 90° from each other in each stacked layer, and combined with 96 steps, this gives us the equation 90 × 96 = 8,640 = 6° 50' 28.8" N.

× 1.5 (3/2) = 1,296

1,296: The Shark Mound on North Bimini Island just east of Florida is at 110° 22' 53.55" W of Giza, which is 129,600. The Pyramid of the Magicians in another ancient Mayan city in Uxmal, Mexico is the only round-cornered pyramid of the Maya. From above it is a square with round angles, implying a circle. It is a combination of rectangular forms containing circular edges. That is squaring the circle in math, which is 360°2 = 129,600 = 120° 54' 20.000" W of Giza.

× 1.333… (4/3) = 1,728

1,728: In Peru, The Great Triangle of Nazca is at 14° 41' 30.104" S = 17,280.

× 1.5 (3/2) = 2,592

2,592: The Bent Pyramid of Dashir near Cairo, Egypt is not collapsing in on itself, it is made up of two separate bodies with two different slope angles. In it we have 8 sides + 8 corners and an apex, which is 9 and give it to 360° and you get 8 × 9 × 360 = 25,920 = 29° 47' 19.01687" N.

× 2 = 5,184

5,184 Believe it or not, the latitude of the Great Triangle in Peru as 17,280 if multiplied by 3, the number associated with a triangle, gives us its longitude West of Giza. 17,280 × 3 = 51,840 = 106° 15'° 32.603" W of Giza.

"I introduced the magnificent code system involving the worlds, pyramids, mounds, maps, and numbers. It all ran the spectrum from simplicity to apparent confusion, as it was supposed to. No one can design a top-secret communication system and leave it right out in the open for everyone to see and trip over, unless it is carefully arranged to confuse the undisciplined mind. And the key to that was numbers."
~ Carl P. Munck[23]

All the Same Numbers

It is astonishing that when you get precise maps of the historical monuments, they start speaking in code, and the language they are speaking is mathematics. Those calculations are devised by many of the same harmonic numbers we see in the solar system, and in our chart of most harmonic numbers in vortex based mathematics by multiplying in ratios of 2 and 3, the doubling circuit going down and the tripling circuit going across. All of these incorporate ratios of 3/2 and 4/3, depending on how close a number is in the chart to its nearest neighbor in the next column, exactly as the circle of fifths neighbor each other in the notes by ratios of 3/2 or 4/3 seen in the diagram, "Musical Notes and Their Hertz Relationship to Each Other." The numbers in the last few charts are not all of the 12 notes of the musical scale, only varying octaves of primarily D, A and E notes. For example, 108, 216, 432, 864 and 1,728 are all an A note; 162, 324, 648, 1,296, 2,592 and 5,184 are all an E note; while 36, 72, 144 and 288 are all a D note. The top three strings on a guitar are tuned to an E, an A and a D. How interesting. As a musician, I can't get enough when it comes to understanding the math of music. It can even be called a science. This truly is the music of the spheres Pythagoras was talking about.

Chapter 6

The Numbers 1 through 9 and Their Numerical Geometry

The Egg Carton Universe

The music of the spheres is showing the expression of harmonic numbers through its vast geometry. It is as if there are invisible lines of force connecting each aspect of its geometry to another, interlocking in a grand symphony and dance of unity. The galaxies themselves are clustered in certain areas, and are separated by unthinkable vast distances between them. There must be an explanation because it is not random. It cannot be random when we start to notice a pattern or geometric arrangement happening. That is what scientists started to discover once they plotted all the clustered galaxies into a grand map of the universe. These clusters of galaxies followed lines of geometry in what some call the egg carton universe. The lattice geometry that is clustering these galaxies together is a matrix of octahedrons from point to point, side-by-side, in a grand web of alignment.[24]

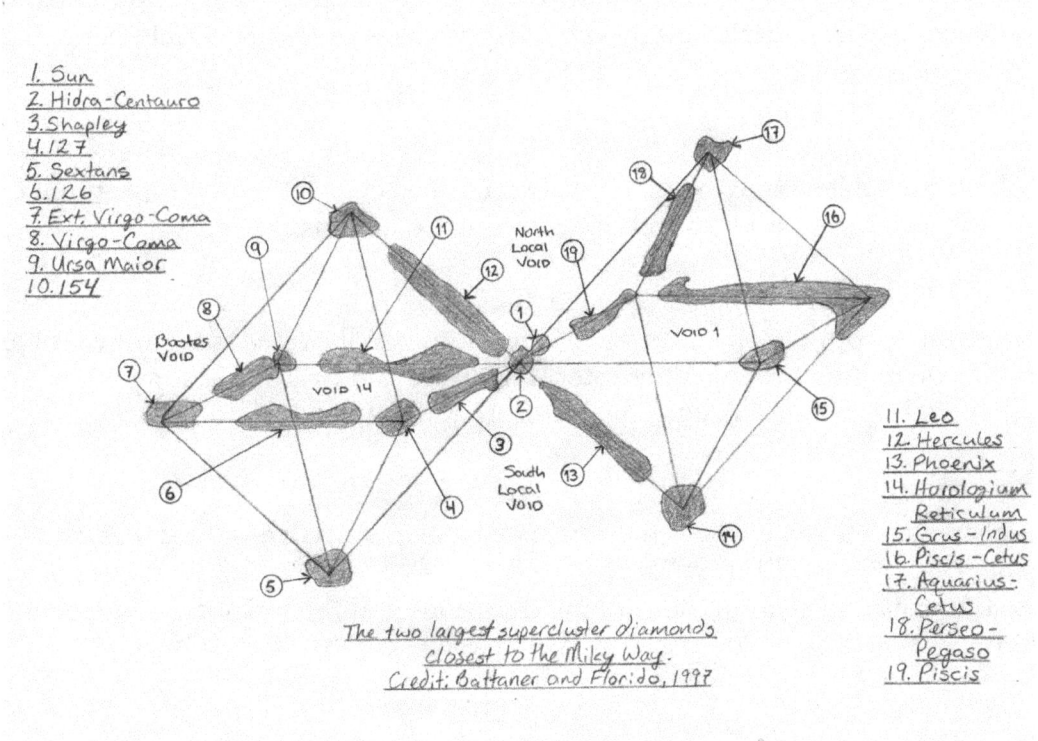

Image 39: Egg Carton Universe

There are many geometries in the creation of form, and it is the numbers that define that geometry. It is consciousness that is behind the numbers. All starts with thought. Like a set of instructions within a system of resonance and frequency, from 9 comes 6 and 3, the beginning of geometry, which does not exist without numbers. So, with harmonic numbers comes harmonic geometry. Numbers as a result must create geometry. Everything, although seemingly separate is still connected as part of one whole working towards unity.

The 360° Circle

Wanting to discover more, I asked myself, "Where can I study these harmonic numbers then?" Within a circle or sphere, of course. I plugged vortex based mathematics and geometry into a 360° circle. I was confident that I would find a correlation, because the patterns, sequences and groupings were so apparent in the numbers themselves. For sure they had to be doing the same thing within a 360° circle.

I started by drawing a circle and breaking it off into 36 points of 10° each. I did it this way because it was either 360 points to calculate, or 36 by reducing each point to every 10°. This gave me every number of 360, just reduced by 10. This seemed like the easier way, but at the same time just as logical. I figured I'd try the simpler method of 36 points before the complex method of 360 points, to see what would happen.

The Numbers 1 and 2

I obviously can't start with one because on a 360° circle, it is either one point or one line. But in two dimensions, which is what these diagrams are in, we find a solution. One line equals 2 halves, 2 lines equals 4 halves, and 4 lines equals 8 halves. A line is required to begin the doubling sequence. One cell breaks into two through a plane in the center, and from the side into two dimensions is a line, but from above it appears as a circle contracting towards the center. Therefore 360° + 180° (a straight line across) is 9 + 9 = 18 = 1 + 8 = 9. A line equals 9 from the point of origin, a straight line from 360°/0° to 180°.

For the number 2, we would have to split the 2 equal halves into 2 to make 4 equal parts. So, from 90° to 270° is 90° + 270° = 9 + 9 = 18 = 1 + 8 = 9. Two lines of course creates the 4 quadrants, as in the cardinal directions and the 4 seasons. So technically 2 would be 360° + 90° + 180° + 270° = 9 + 9 + 9 + 9 = 36 = 3 + 6 = 9.

The Numbers 3 and 6

The numbers 1 and 2 are the simplest, but initial numbers within the 360° circle need a third number to create the first shape possible as a polygon, a triangle, of course, with 3 points. This is the test I have been waiting for. Getting all 9s, with 1 and 2, at least in their position of origin, was pretty cool, but this should get better. I started with the top of a triangle at 90° with its first point, calculated 120° to the next point (360° ÷ 3 = 120°), which was 210° and another 120° to the next one at 330°. I took the root sum of the triangle and it gave me 936. That is exactly what we saw in the 24-digit Phi sequence of the 12-digit half containing the 3's, 6's and 9's. The triangles in my Trinity of Creation diagram were composed of two triangles connecting 3, 6, and 9, one triangle connecting three 1's, and another triangle connecting three 8's. So, within that diagram, the prominent numbers for a triangle were 3, 6, and 9. Perfect, it checks out so far! But will vortex based mathematics continue to surprise me the further I go with this method? On the next page are the 6 combinations from a 36 point circle.

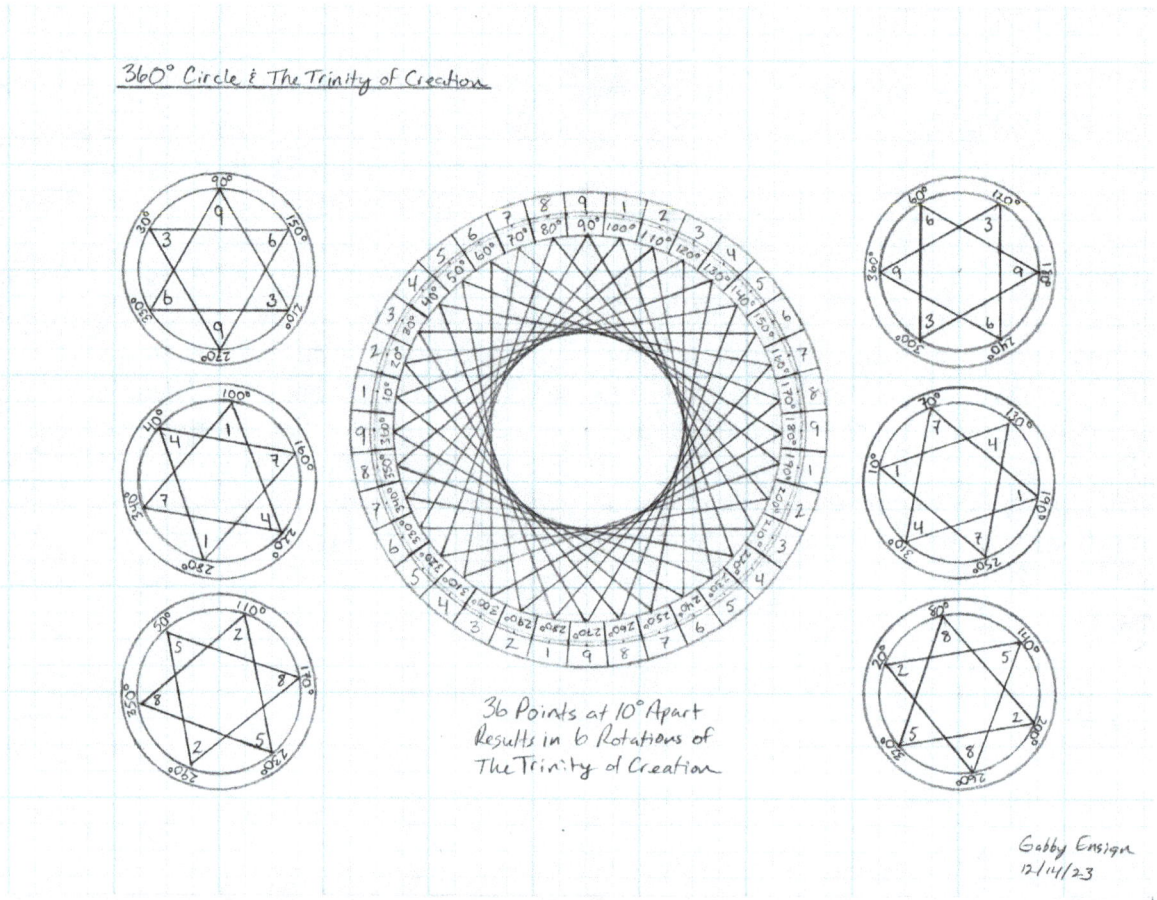

    I did not create this diagram. Along with the others I discovered it. An equilateral triangle in a 360° circle will always tell us the relationships between the numbers and their functions. It shows us that through the geometry of an equilateral triangle within a 360° circle, it will give us the 396, 147 and 852 family number groups. This blew me away. It showed me that even the simplest geometry in a 360° circle contains a very fundamental aspect of vortex based mathematics and of sacred geometry. My test was a success. I call this diagram, "360° Circle and the Trinity of Creation." Just as the Trinity in many ancient cultures works through the higher dimensions to help facilitate the spiritual within the physical, so too does the trinity of 396. They are etheric and function in a higher dimension to create the physical.

    This diagram also shows us that 3 must exist within 6, because the balance of 3 is its opposite, to create a 6 pointed star tetrahedron, a Hebrew symbol known as The Star of David. In geometric terms, it is a star tetrahedron, one of the most fundamental forms we see in nature through the form of a hexagon. It also shows us that 9 is still a function working with 3 and 6, but 9 being the source of all function, works as the balance for 3 and 6, and all other numbers created within the system.

The Numbers 4 and 8

Just like one line creates two halves and 4 is a function of 2 (as in the cardinal directions), 8 is a function of 4, because the square doubled is an octagram. Within this next diagram, you can see 4 in its geometry and numerical sequence, and 8 becoming its completion. The number 4 is a square in two dimensions and becomes 8 in three dimensions making a cube. So, in this two dimensional representation of 4 in vortex based mathematics, it shows 8 in its completion with 4 because with 8 numbers, 4 are even and 4 are odd. The number 8 appears to need the duality of 4 to be whole in its function. Therefore, the numbers 4 and 8 are combined into one graph I call, "360° Circle and the Octagram."

In this next diagram, I was determined to find more than what I initially found when plugging 8 points equally spaced within the 360° circle and aligning to 0°/360°. 360° became 9, 45° became 9, 90° became 9, 135° became 9, and so on. I couldn't see anything other than 9. So, I figured I'd try thinking outside the box. Like Carl Munck said, codes can be hiding right out there in the open, but arranged to confuse the undisciplined mind. OK then, let's go by the ratios of each point within the circle, I thought. 360°/360° = 1, 45°/360° = 1/8 = .125 = 8, 90°/360° = 2/8 = .25 = 7, 135°/360° = 3/8 = .375 = 15 = 6, and so on. Check out the diagram on the next page to see the number sequences and their geometries.

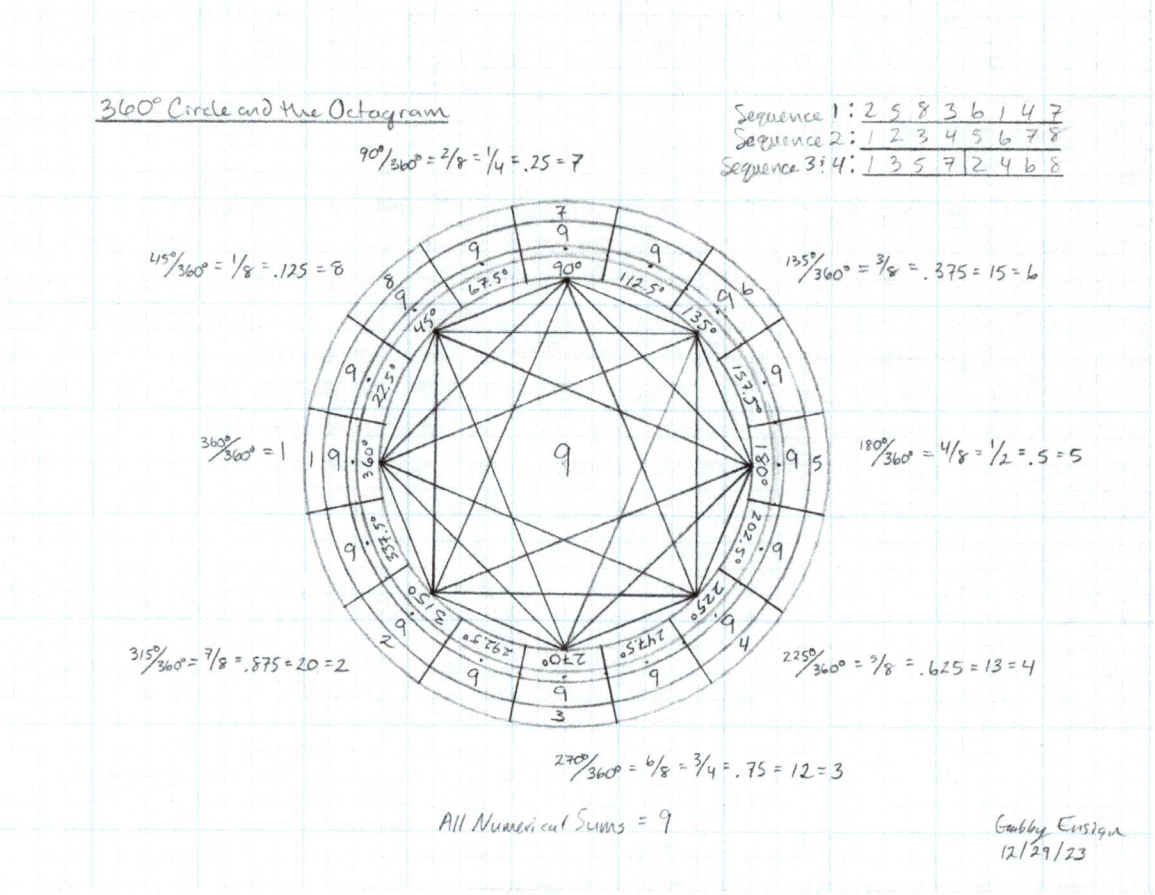

Right away, I noticed 3 familiar sequences in the 8 digits circling the 360° circle. The first one was very obvious, which is 1 2 3 4 5 6 7 8. The second, instead of moving to the next point around the circle, would move to every other point. This of course, only completes one square of 1 3 5 7, the other one in a separate sequence is 2 4 6 8, even and odd in 2 squares to become the full sequence of 1 3 5 7 2 4 6 8. The third sequence is every third point, which like the first, reaches all 8 points in the circle, very similar to the Diamond of Fifths patterned sequence.

This forms the octagon, it reveals the final of the 3 sequences revealed to be 1 4 7 2 5 8 3 6. There are again 3 distinct sequences to make this pattern complete. Two of the three are exact, but the third only follows a similar pattern because we are still missing 9, although it is the 3 family number groups side by side, minus the 9. So other than a slight variation, it is Marko Rodin's toroid number grid maps of the 147 Shears, running at 45°, and the same ones in Gator's toroid number grid maps, also running at 45°. These numerical frequencies hold the inner and outer construction of its corresponding geometry in vortex based mathematics. The 360° circle and the octagram are now established as an expression of vortex based mathematics. The number 9 is the still center, the point of unity for the whole. And astonishingly, my hypothesis was correct in finding the ratios of each point as it relates to the whole.

The Number 5

If 4 and 8 were getting close to the complete sequences of Rodin and Gator's grid maps, maybe the number 5 would contain them all. With every number so far from 1 through 4 increasingly showing me more and more, as if each one carries one more level of complexity as it progresses to the next number, then the number 5 should hopefully follow suit and show me something interesting and valuable.

I drew out my 360° circle and plotted 5 equally distant points starting at 0°/360°, 72°, 144°, 216° and 288°. All equal 9. Ok, then, I will apply the same logic, but first I thought there needs to be more than one pentagon, and like the others would require at least two. I came to a standstill. 360° divided by 5 equals 72. So that is 72 pentagons in this circle. I was sure that would show me something, but I thought there could be another way to divide. I pondered some numbers and calculations, but decided to put it down for the night and pick up where I left off the next day. I would have to sleep on it.

Later the next morning after breakfast and coffee, I didn't want to jump straight into all of this right away and I still needed a break, so I watched a little TV. I found an interesting program on about the Knights Templar and the cathedral-like structures they built. It looked interesting enough, so I continued to watch for a while. Then I saw an image of a large cathedral with a 12 sided stained glass window in a beautiful mandala pattern. Yes, I actually counted the petals. It didn't catch me off guard much because I knew the Templars displayed 12 in their construction for a long time. Then I saw a pentagram inside a pentagon inside a circle, with three rays extending outward between each of the five spires of the pentagram, which is mounted above on an altar or window in a small sanctuary. I was not sure where this sanctuary was located, but that didn't matter. It gave me an idea.

I went back to my circle with one pentagon and said, "Why not?" I had a good feeling and the timing around seeing this image was a little weird, so I decided to run with it. Right away a bunch of numbers started popping out. I divided 360° by 5 to get 72° like before and then I divided again by 4 for 1 pentagram with 3 spires between angles which equals 4 pentagrams or 4 pentagons. The equation in a more simplified form is $360° \div 5 = 72 \div 4 = 18° \times 20$ equals 360°. That is 20 equal points apart by 18°. The number 18 is a harmonic vortex based mathematics number with the 18-digit sequence of the Rodin 147 Shears and is 9 × 2. Look at the innermost set of numbers, indicating the degree of each of the 20 points in the diagram. Some are identified with a ● symbol.

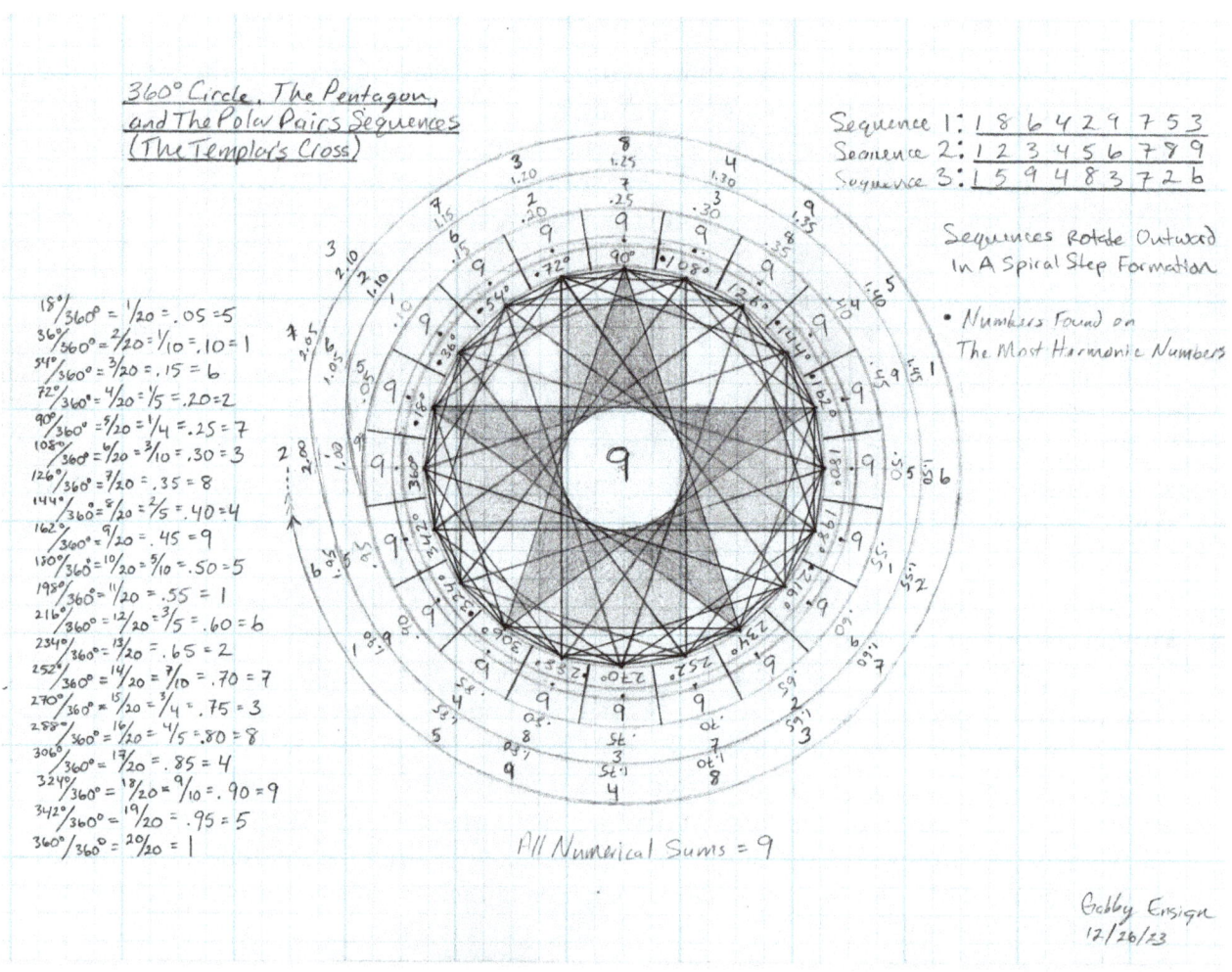

The majority of them are listed in the chart of most harmonic numbers. They all have a root sum of 9, which is expected at this point, considering that every number in the chart of most harmonic numbers right of the second column has a root sum equal to 9. Even the very long decimal numbers equal 9. Well, except I rounded up on some, because I couldn't fit all the digits in the cell, but I assure you they all do.

The next step was to divide each point to its relationship to the whole. So, 18°/360° = 1/20 = .05 = 5; 36°/360° = 2/20 = 1/10 = .1 = 1; 54°/360°= 3/20 = .15 = 6 and so on. The diagram contains all of the calculations.

You can see that it is spiraling outward in its sequence. I was a bit confused the first time I calculated the number association for each point because the sequences of numbers that I was getting were so close but would fall off very quickly. Since each point in its ratio equation increased by .05 each time, the next step past 360° at 1.00 to 1.05 would increase the sum of the next point by a value of one more than the previous in an outward spiral. Once this is done the sequences all line up. We can see a reoccurring pattern has emerged now as the three 9 digit polar pairs sequences also seen in the Rodin and Gator grid maps, 1 2 3 4 5 6 7 8 9, 1 3 5 7 9 2 4 6 8 and 1 5 9 4 8 3 7 2 6. They are spiraling outward and inward just like a toroidal vortex does in its flow of movement. This will become relevant again in Chapter 8.

The sequences aren't difficult to find, but can be tricky at first. The first one is easy. It is the sequence that goes sequentially in line around the spiral starting at 18° which is .05 or 5 then to 1 6 2 7 3 8 4 9 and back around to 5 to repeat the sequence. Did you notice that the sequence shifted up from the 5 to 6 at 18°? This is showing us the 1 2 3 4 5 6 7 8 9 sequence, but if we use the logic of instead of every next point to every other point, we get again the sequential pattern of 1 2 3 4 5 6 7 8 9. And, 18° to 54° is a step by two points, and 18° = 5 and 54° = 6. From 54° it goes to 90° = 7 and so on every other one. So, there are two ways to get to 1 2 3 4 5 6 7 8 9. The sequence above as 5 1 6 2 7 3 8 4 9 I would usually write as 1 5 9 4 8 3 7 2 6 because it starts on one and goes to its fifth around the sequence of 9 in most of my diagrams.

The other location of the sequence is in the five pointed star pentagram pattern in the middle. The points are in sequence to how you would draw it on paper; 18° = 5 which goes to 162° = 9; to 306° = 4; 90° = 8 because it is the next number in the sequence, considering going from 306° = 4 back around to 90° you pass 360° and would then go up one in a step around the spiral. From 90° = 8 it goes to 234° = 3; back to 18° but is a 7 because again passing 360° in the sequence brings it up from 18° = 6 to 18° = 7. The sequence 5 9 4 8 3 7 2 6 1 continues spiraling outward to inward to infinity. Again, I would see this as 1 5 9 4 8 3 7 2 6 just for consistency.

The third sequence of 1 3 5 7 9 2 4 6 8 could be written as 1 8 6 4 2 9 7 5 3 if seeing it in reverse and again starting with the 1, as it is written in the diagram to line up other numbers. This one is a little harder to find but makes total sense. Applying the shape of a pentagon instead of a pentagram, we would move to every fourth point. 18° = 5 goes to 90° = 7 to 162° = 9; 234° = 2; 306° = 4; 18° = 6 because it has now passed 360° and 18° is now 6, up from 5; then to 90° = 8; 162° = 1, and so on around the spiral, 5 7 9 2 4 6 8 1 3, which again is written on the diagram as 1 8 6 4 2 9 7 5 3 just to lineup numbers. This sequence I would personally see as 1 3 5 7 9 2 4 6 8 for simplicity and recognition. I only see this sequence once, whereas I see the other 2 twice. I am not sure why, when I was expecting to see two like the others. Maybe it is supposed to be that way for some reason, like 2 + 2 + 1 = 5; 2 sequences plus 2 sequences plus 1 = 5 sequences. Or maybe not, it is all just observation at certain points. I may not be seeing them all.

Well, 5 sure did one up in its complexity from 4 and 8. Just as expected. But why do I also call the diagram the Templar's Cross? Do you see the cross in the lighter shaded areas? It is the cross from the Templar's church which I saw on the documentary. The "hint" that I got by putting my project aside to give the proper solution time to present itself was a good decision. I just didn't think it would happen the way it did, but it did. Aren't synchronicities amazing?

The Number 7

After the number 5 is 6, but 6 was included with 3 and its diagram so that makes 7 the next of 9 numbers to look at. I don't know if I'm ready for 7 after the elegant complexity of 5. But my curiosity couldn't keep me from finding out. I discovered 5 the day after Christmas and a lot was going on so I couldn't get to 7 until a couple of days later. I could hardly wait actually; despite the challenges I might run into with 7.

Starting the same way like all of the rest, with a 360° circle, I plotted 7 points on a circle, which was not too difficult; 360° divided into 7 points separated by 51.4285714285714...°, an infinite number with only numbers from the doubling sequence. And double that is 102.8571428571428...° into infinity. The same numerical sequence in the decimals is the same as before, except it is shifted slightly. I confirmed with the next number, and the same held true to all the rest, the decimals repeated the same sequence of numbers.

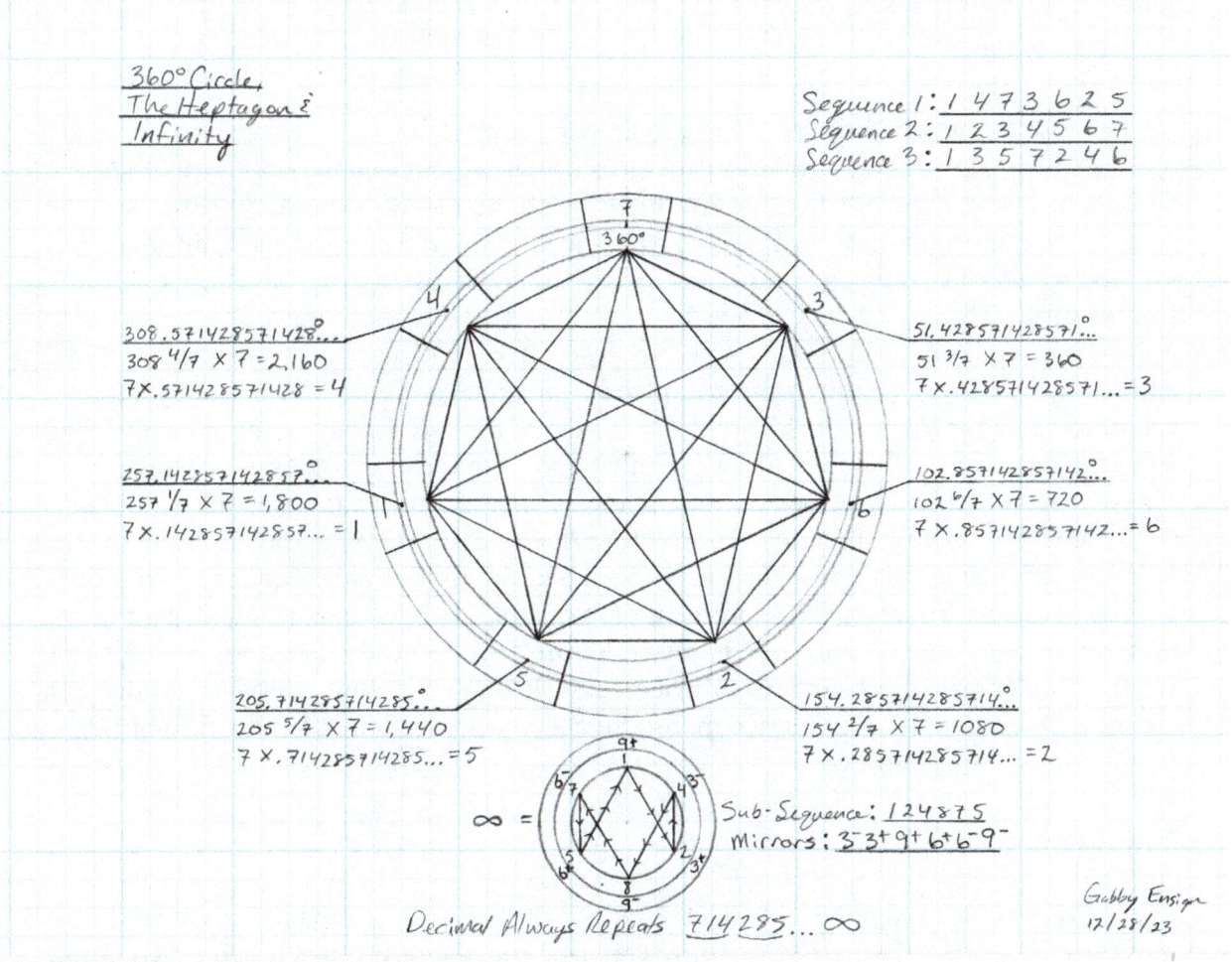

I then converted the decimals to a fraction to see if that made things easier. Sure enough, .428571428571 is exactly 3/7. So, our first point is 51 3/7. Going around the circle I did the same. When I got to the last one, I started to see a small pattern. Each one of the fractions was a fraction of 7. So, I assigned each point from the numerator of their fraction; 51 3/7 became 3 because the number said it was 3 of 7. Then 102.6/7 becomes 6 and so on. This gives us the sequence 3 6 2 5 1 4 7. I did not recognize it so I felt it was not correct in my method. But then it clicked. Starting with 1 then going every other one while looping around gave me the sequence 1 2 3 4 5 6 7. OK then, this might be right. I attempted to find 1 3 5 7 2 4 6, the odd then even sequence and it was the same method to get the other star patterns like in the Rodin 147 Shears diagram and the Pentagram of 5. I guess this works out.

But why the outlier in the first sequence? I looked at it a minute and saw that in the sequence, the numbers 1, 4 and 7 were grouped, 3 and 6 were together, and 2 and 5, (which is minus the 8 in the family number group 825) were together. At least this outlier sequence makes some sense in vortex based mathematics. Can I trip it up on the next attempt?

I do not know why I did not think to take the root sum of the whole number along with its decimal number in fraction form to find a value, because on every other one I tried that first. I guess since I started getting 9's across-the-board with the octagon and the pentagon and had to find a clever way which was using ratios, I automatically jumped to just assigning the numerator of the fraction instead of the whole root sum. For example, 51 3/7 = 5 + 1 + 3 + 7 = 16 = 1 + 6 = 7. Next, 102 6/7= 1 + 0 + 2 + 6 + 7 = 16 = 1 + 6 = 7. 154 2/7 = 1 + 5 + 4 + 2 + 7 = 19 = 1 + 9 = 10 = 1 + 0 = 1 and so on. We now get something completely unexpected, in fact more interesting than the first outlier sequence. It is 7 7 1 1 4 4 9. Again, the 147 family number group! And it is accompanied by 9 with 360°. Wow, this tells me that the first method of taking the numerator to assign a number in the circle was correct for what I was doing, but this sequence does not contain any number other than 9 and the 147 family number group. It omits 825, just as it is not complete without 8 in the first outlier sequence. But it does give us some information. The root sum of the angle with its decimal equivalent in the number 7 says, "I am 7 and here is my function with 1 and 4." I cannot trip this one up. It is vortex based mathematics through and through. The only thing left to do with 7 is to take a closer look at its decimal sequence because it is not 1 2 4 8 7 5, but does contain only those numbers.

The Decimal of 7

In the previous diagram, you can see at the bottom a smaller circle with the numbers 7 1 4 2 8 5. Is there something very obvious about it now? It is the two family number groups side-by-side. I did not recognize it when writing the decimals down on paper, but I got it when I drew this circle. Now, what do I do with it? Of course, find its relationship with 1 2 4 8 7 5. As you can see it is infinity in a hexagon, and a hexagon is the polygon of 3 and 6 with the two interlocking triangles making up the star tetrahedron. Why am I mentioning this? Because the 1 2 4 8 7 5 pattern that is showing up in the mini diagram of the 7 1 4 2 8 5 family number group sequence is the 3– 3+ 9+ 6+ 6– 9– path of polarity. You can see it with the + and − around the outer circle.

This back-and-forth, flipping and flopping of polarities is kind of like a wave pattern. When seen through the functions of 396, we see it as a path, more than a wave, because a wave is physical and would be within the domain of 1 2 4 8 7 5. That is why the 3 and 6 are not connected except through the 9 in the Rodin Symbol.

Through 1 2 4 8 7 5 operating within the 147 and 852 number groups, it would represent something non-etheric and more physical in our universe. That being said, since 1 2 4 8 7 5 is a 6-digit sequence and 6 is etheric, 1 2 4 8 7 5 is operating and being directed by 6. And, since 1 2 4 8 7 5 is two family number groups of 3, 1 2 4 8 7 5 is operating and being directed also by 3. Therefore, 1 2 4 8 7 5 is operating and being directed by 3 and 6, and they are of 9, so it resonates with the 3− 3+ 9+ 6+ 6− 9− circuit within the mini diagram of 7 1 4 2 8 5.

At this point, I could add 7's diagram to the binder of vortex based mathematic sitting on my desk. It was about two weeks later when I decided to look at the mini diagram of 7 1 4 2 8 5 again and play with its inherit harmonic design because I knew it would speak vortex based mathematics. The next diagram shows the harmonic relationships between any opposing number within the diagram. The middle circle of 7 is the diagram itself with no adjustments. The other 6 are the possible combinations of opposite relationships. Around all of the 7 circles are an inner and outer set of numbers. The inner is the original sequence, and the outer is each numbers opposite of its given point. The rest you can get from there, so I will let the diagram speak for itself.

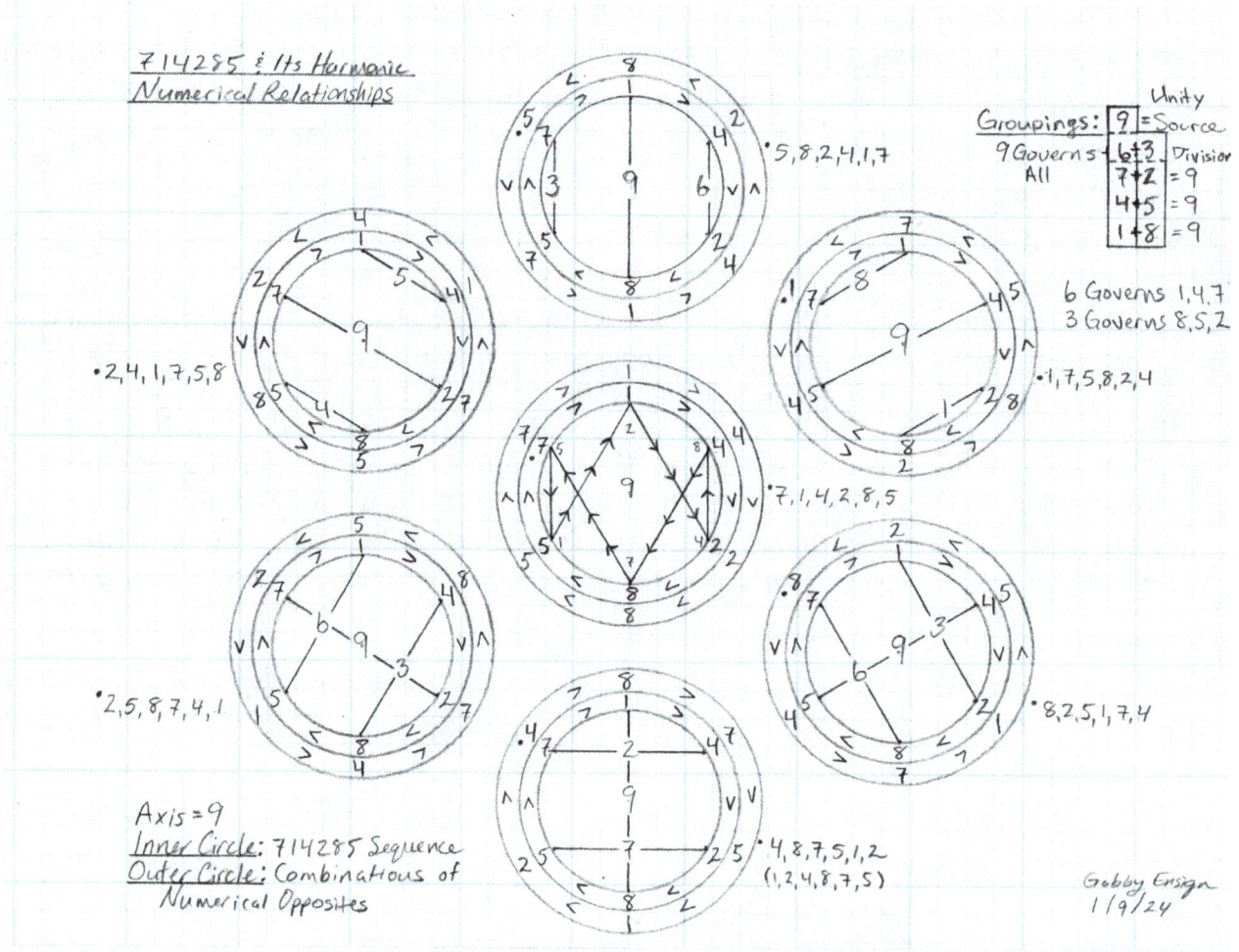

Angle and Ratio

> *"Numbers are relatives. Each number exists only in relationship to all other numbers. The relationship is spatial and temporal, which is angle and ratio."*
> ~Marko Rodin[25]

My experiment with vortex based mathematics within a 360° circle was a clear expression of angle and ratio. Every point had an angle and every point was defined by its ratio. I did not draw a diagram for 1 or 2 but 1 would have an angle of 180° and a ratio of 1/2, while 2 would have an angle of 90° and a ratio of 1/4 (360° ÷ 2 lines = 4 quadrants) = 90° and 360°/4 = 90°. So, 3 and 6 show the root sum of their point angles at 60° apart and carry a ratio of 1/3, 2/3, and 3/3. Because the 3 3 9 6 6 9 circuit progresses by 3, there are 3 different numbers in the 396 family number group, and the 3 6 diagram is made up of two 3 sided pyramids in a whole of 3/3 to complete the trinity of 369. As always, 9 is close by to 3 and 6. Also, 4 and 8 have points defined by their angle and their ratio with 360° and equated to their root sum to assign a number value to each point. The same goes for 5 with its relationship to the 20 points at 18°, plus it is unique in its complexity; the number 7 is very unique in its complexity, maybe even more so than 5, but at the very least different.

The Number 9

Now that we have 1 through 8 plotted within a circle, all we have left is 9. The number 1 gave us 360° and 180° which are root sums equaling 9, and 2 gave us that with 90° and 270° (both root sums of 9), 4 and 8 gave us the angles of 1 and 2, plus 45°, 135°, 225° and 315° (which are all root sums of 9), along with the number 5 having points of a root sum of 9. The number 7 is an outlier because the root sum of its whole numbers are 1, 7, 5, 2, 3, 6, and 9. The number 7 is showing characteristics of 3 and 6 and seems kind of special in that regard, like a number that can walk in both worlds of the etheric and the physical. Funny, that is exactly what 7's mini diagram showed in 7 1 4 2 8 5.

So, considering 3 and 6 defined their points as just described, I assume 9 will do the same thing. But I cannot make any assumptions because 9 is the most important number. Either way it has much to show us.

The following diagram shows the enneagram of 9 in its 4 combinations within a 360° circle at points 10° apart. That is 360 ÷ 9 = 40 and 40 ÷ 10° = 4. Not surprisingly we see the continuance of the polar pair sequences and with the full geometry of 9 within a 360° circle. It is beautiful music to my eyes.

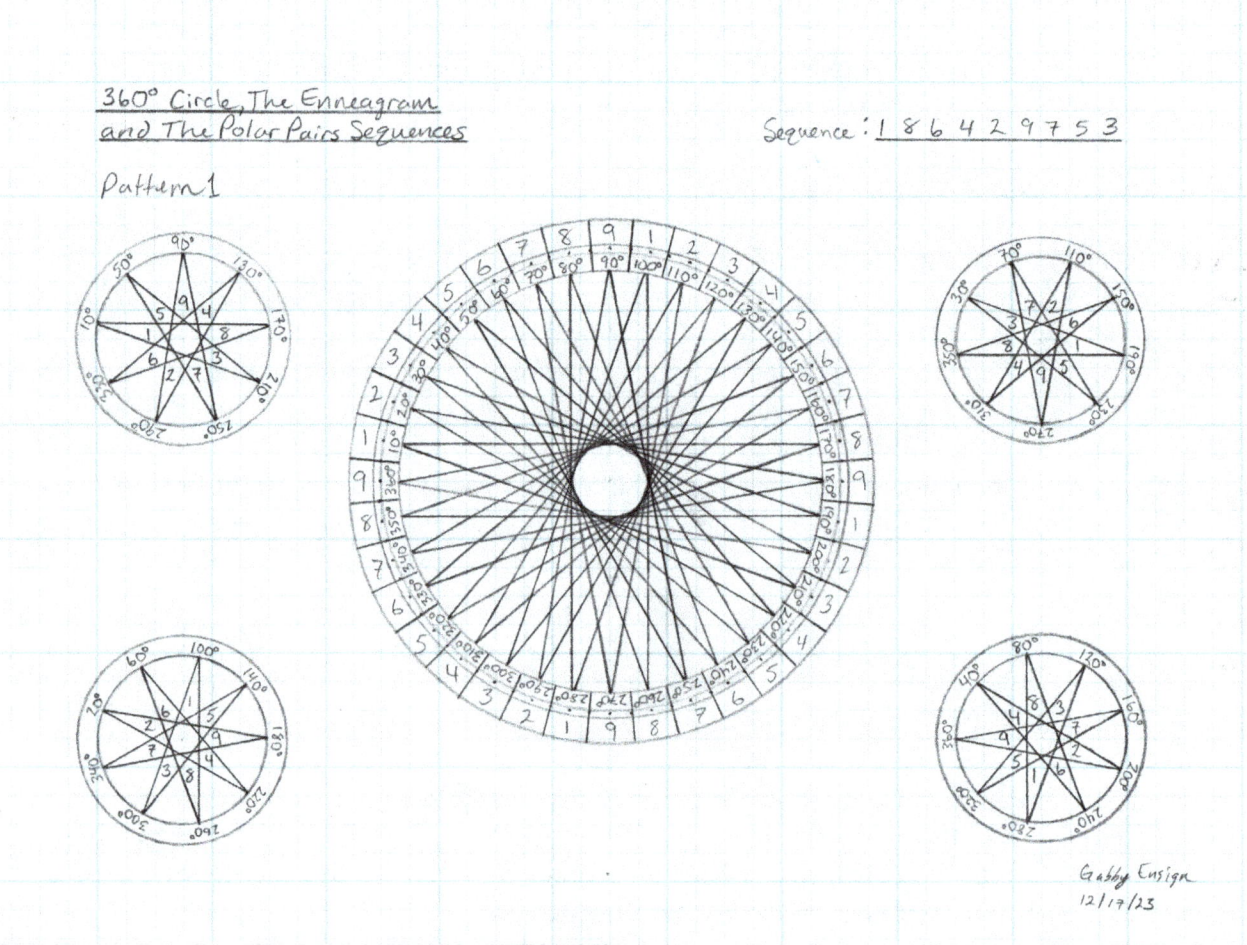

So, what have our assumptions given us? We are correct in our assumption. that like 3 and 6, the points of 9 on the 360° circle get their value from the root sum of each angle of 10° instead of that points value originating from a calculation of its ratio to the whole. The sequences of the four different enneagram combinations gives us a sequence we have been looking for. It is as mentioned, the 1 8 6 4 2 9 7 5 3 sequence of 9 numbers at 45° within the Rodin 147 Shears, and completes the previous 4, 8, 7 and 5 number sequences in their attempt to move towards all three polar pair sequences of 9 digits. The number 9 gives us a new set of geometry within 36 points. Pattern 1 shows us that there are 40 paths to 9 in a 360° circle and since we have divided the circle into 36 points of 10°, we get 4 paths to 9 within a 360° circle at 10°. This pattern is one of three different patterns, showing us the most balanced, efficient and direct ways to hit each point in a circle of 9. This pattern again resembles the "star" pattern seen in all of the other polygons so far in one form or another, depending on their geometry. Such a beautiful wheel of spires that look a little bit like petals on a flower.

360° Circles, The Enneagram and The Polar Pairs Sequences

Pattern 2

Sequence: 1 2 3 4 5 6 7 8 9

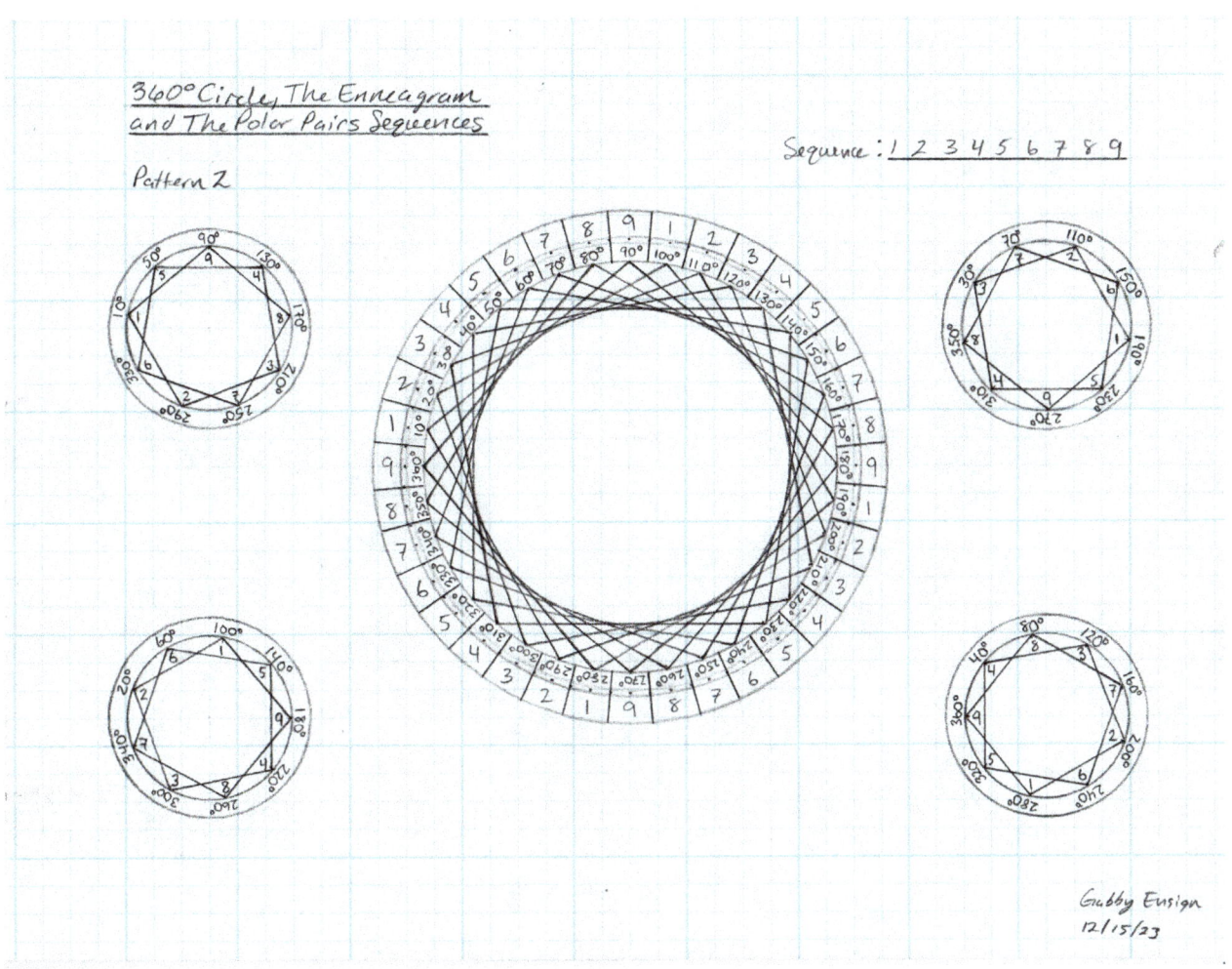

Gubby Ensign
12/15/23

360° Circle, The Enneagram
and The Polar Pairs Sequences

Sequence: 1 5 9 4 8 3 7 2 6

Pattern 3

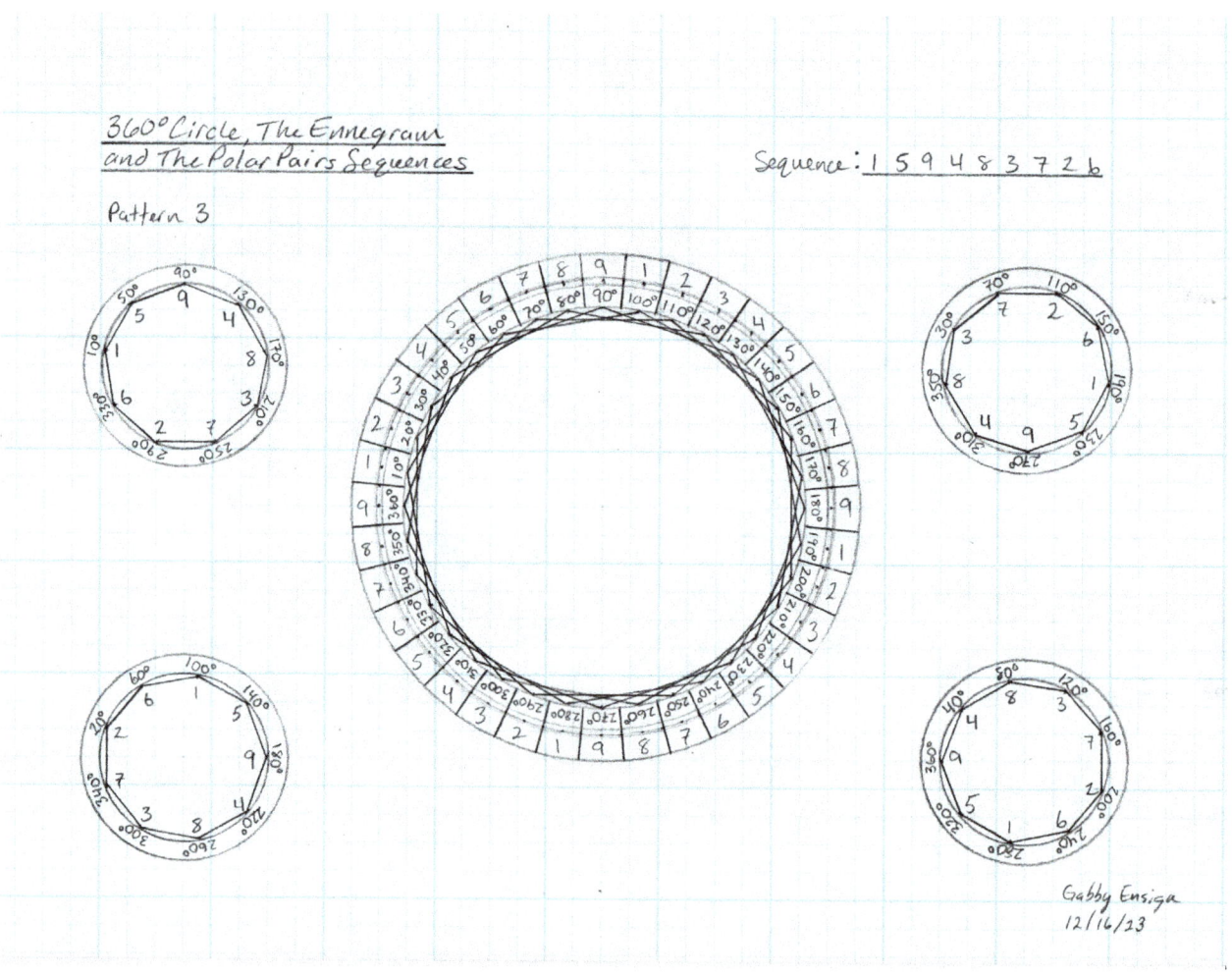

Gabby Ensign
12/16/23

Patterns 2 and 3 give us the two other polar pair sequences we have been looking for. Pattern 2 skips every other point of each of the four different combinations of 9 with a different starting point of 1 and a different ending point of 9 landing at each of the four cardinal directions of 360°, 90°, 180° and 270°. The sequence in pattern 2 is 1 2 3 4 5 6 7 8 9.

Pattern 3 is the numerical sequence pattern from one point to the next closest point usually seen as 1 2 3 4 5 6 7 8 9 going around a circle, but this time it is giving us the last polar pair sequence which is 1 5 9 4 8 3 7 2 6, which is the sequence that would usually give us the "star" pattern with a sequential numerical sequence of 1 2 3 4 5 6 7 8 9 going around the circle.

So, there we have it, all 3 sequences of the polar pairs in the 360° circle enneagram diagram. They all perfectly correspond and mirror each other the same way that 4, 8, 7 and 5 do in the 360° circle.

The Enneagram and the Trinity of Creation

The number 9, then embodies, both simplicity and complexity as it contains all aspects of the whole. These next two diagrams show us this in a beautiful mandala of balanced symmetry. The first one containing just the three polar pairs sequences, and the second containing, in addition, the "360° Circle and the Trinity of Creation" diagram, which I added to complete the apparent void in the Mandela. It seemed like the logical thing to do. Take a look at the second diagram. Isn't it mesmerizing?

The complete structure of the geometry of 9 within a 360° circle is incomplete without the geometry of 3 and 6. The geometry of the polar pairs rely on 3 and 6 to complete the balance of all of the geometries. Numbers 3, 9, and 6 merge with 1 2 4 8 7 5 to create its full form. Since 9 is "All" in vortex based mathematics, the invisible point of unity, the shear point, and the source of all movement, form, and energy in the universe, 9 balances 3 and 6 with 1, 2, 4, 8, 7 and 5 through a point of origin; the center where 9 actually originates from. It gives us a 3 layered trinity of complexity getting ever more complex as the center of 9 moves outward from the center of the circle. We could almost see this as 3 nested toroids. I could stare at this mandala of geometry for hours, and could even imagine it as a stained glass window full of many different bright colors shining in the light that is brightening up a room.

360° Circle, The Ennegram
and The Polar Pair Sequences

Patterns 1, 2 & 3

Groupings: 3 6 9
           2 5 8
           1 4 7

Sequence 1: 1 8 6 4 2 9 7 5 3
Sequence 2: 1 2 3 4 5 6 7 8 9
Sequence 3: 1 5 9 4 8 3 7 2 6

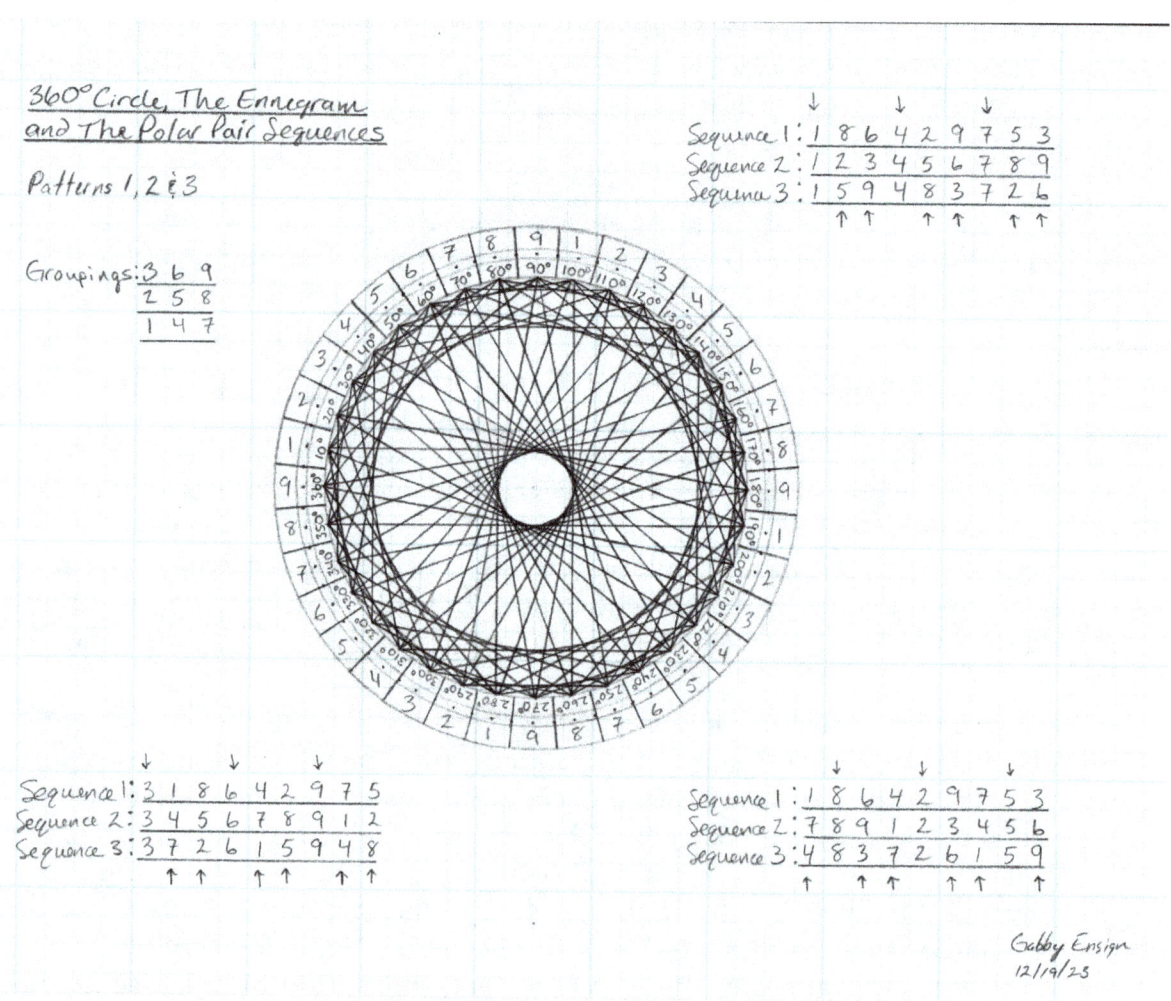

Sequence 1: 3 1 8 6 4 2 9 7 5
Sequence 2: 3 4 5 6 7 8 9 1 2
Sequence 3: 3 7 2 6 1 5 9 4 8

Sequence 1: 1 8 6 4 2 9 7 5 3
Sequence 2: 7 8 9 1 2 3 4 5 6
Sequence 3: 4 8 3 7 2 6 1 5 9

Gabby Ensign
12/19/23

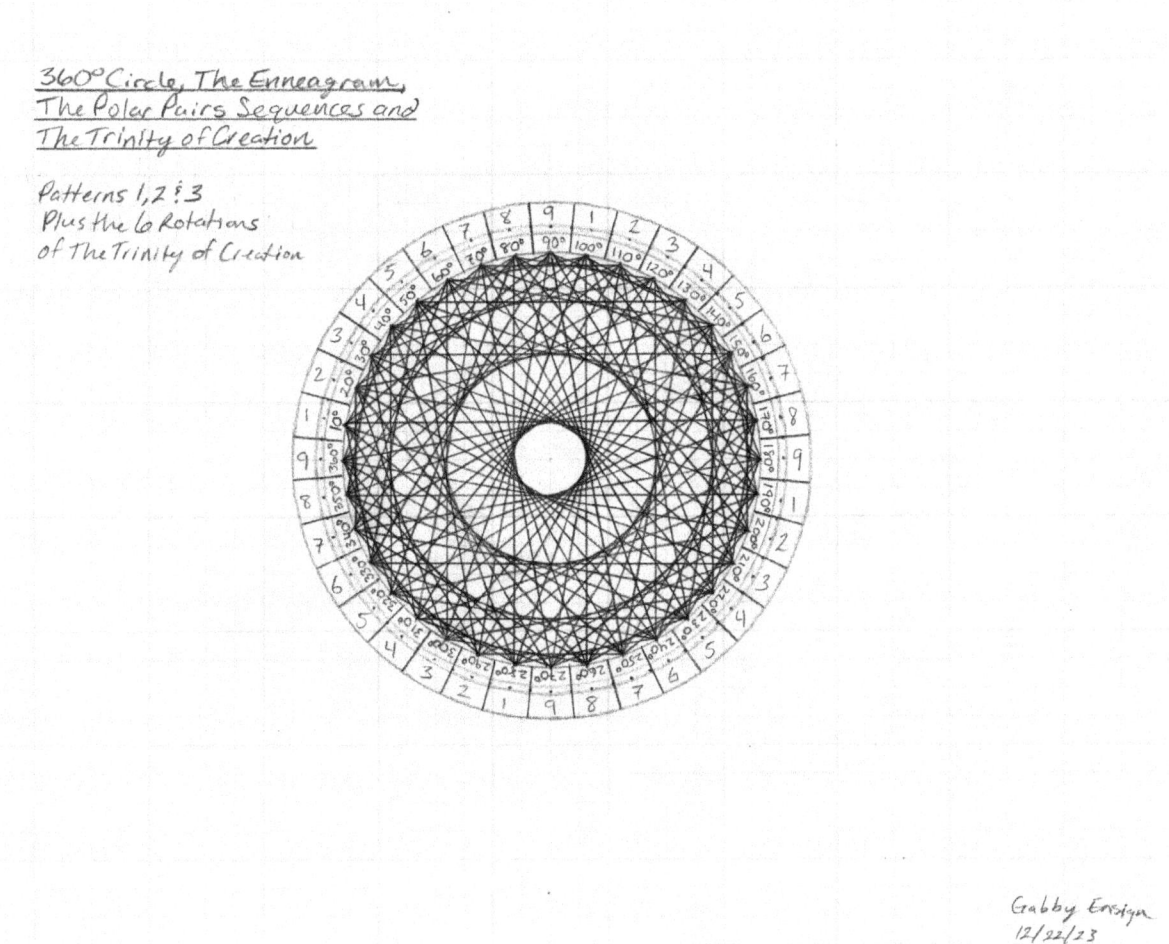

Harmonic Alignments

The three sequences, if aligned properly to the number 1, show harmonic alignments with the three family number groups. There are three ways to align the three sequences to get the three different family number groups of 147, 852 and 396. See the following diagram. The first set of three is lined up to the number 1 which results in 4 and 7 also having triple digits, indicated by an arrow above the sequence and within circles below. The second set displays an alignment of 3, 6 and 9 and the third with 8, 2 and 5. It is crazy how these number groups keep popping up. Maybe Nikola Tesla was onto something with his 369 key to the universe thing going on. He did discover the AC in AC/DC.

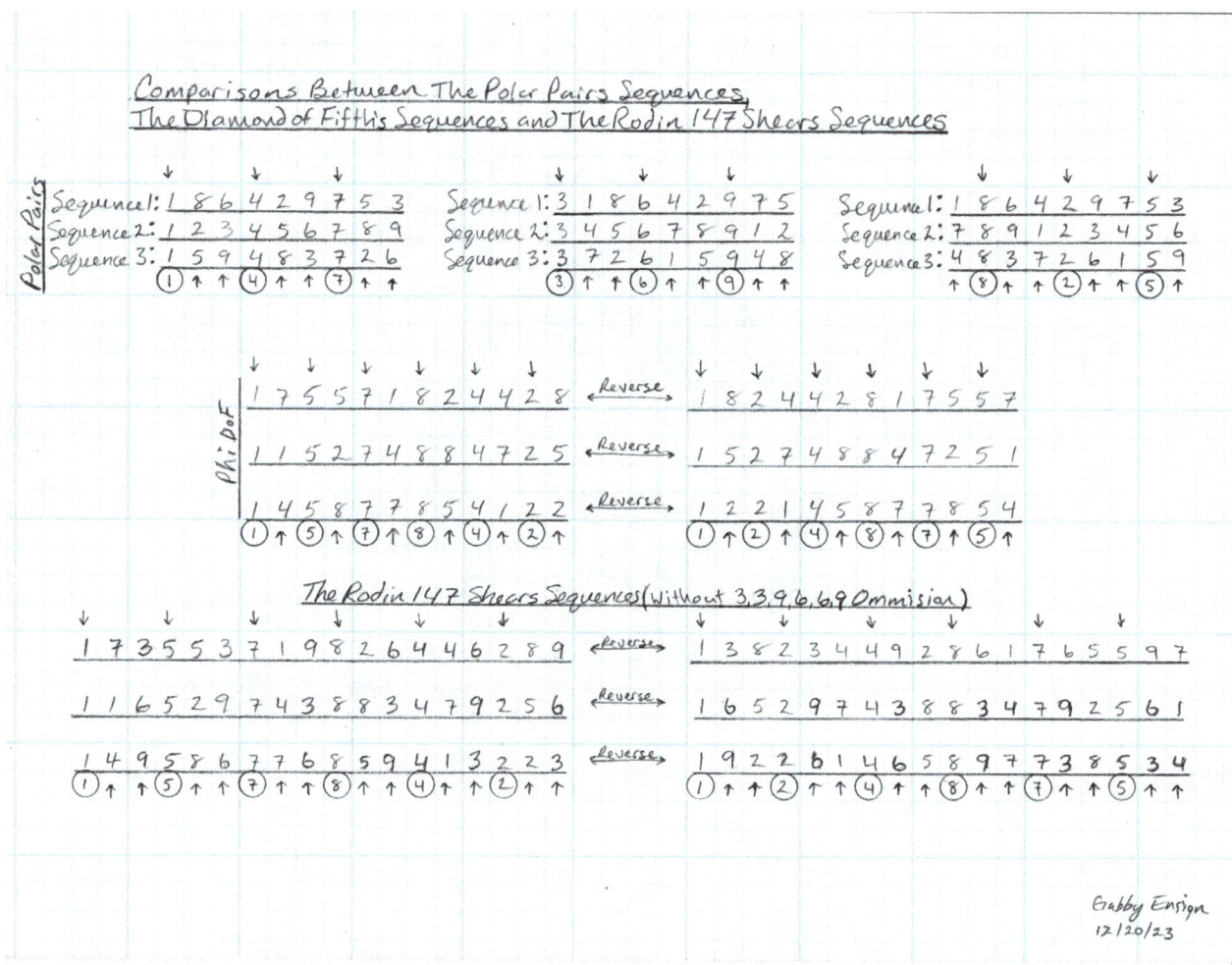

I was starting to see the same family number groups in the other 18-digit sequences of Rodin's 147 Shears and Gator's number map. The Diamond of Fifths sequence is Rodin's and Gator's 18-digit sequences minus the 3 3 9 6 6 9 circuit (18 - 6 = 12-digit Phi Diamond of Fifths sequence shown in the diagram). The same number groups of 147 and 852 appear, along with the 396 in Rodin's and Gator's, and a 1 2 4 8 7 5 doubling sequence interwoven within the circled numbers of three of a kind. In vortex based mathematics terms, the way that it all stays true is kind of mind-boggling. This comparison diagram tells it all.

Chapter 7

The Number 9 - The Point of Balance in a Bipolar Universe

The Axis of 9

"The world naval is the symbol of the continuous creation: the mystery of the maintenance of the world through that continuous miracle of vivification which dwells within all things.

"The dome of Heaven rests on the quarters of the Earth.

"Hence the traditional importance of the mathematical problem of the quadrature of the circle: it contains the secret of the heavenly into earthly forms.

"The hearth and the home, the altar and the temple, is the hub of the wheel of the Earth, the womb of the universal mother whose fire is the fire of life.

"And the opening of the top of the lodge — or the crown, pinnacle or lantern of the dome — is the hub or midpoint of the sky, the sun door through which souls pass back from time to eternity, like the savor of the offerings burned in the fire of life, and lifted on the axis of ascending smoke from the hub of the earthly to that of the celestial wheel.

Thus filled, the sun is the eating bowl of God, an exhaustible grail abundant with a substance of the sacrifice whose flesh is made, indeed, and his blood is drink, indeed.

"At the same time, it is the nourisher of mankind. The solar ray igniting the hearth, symbolizes the communication of divine energy of the womb of the world — and is again the axis uniting, and turning the two wheels. Through the sun door the circulation of energy is continuous. God descends, and man ascends through it.

"I am the door, by me if any men enter in, he shall be saved, and shall go in and out and find pasture."
~Joseph Campbell, "Hero With a Thousand Faces" [26]

Such an elegant piece of writing by the author of "The Hero's Journey," Joseph Campbell. And so appropriate in context to a 9 being the source of all life, the axis in which all things turn, and the opening of the dome bridging the world of the heavens to that of man. The wheels in this quote could be analogous to the wheels of the enneagram, where 9 is the source behind all, in a trinity of celestial and earthly movements through time. Like the north and the south poles, the reciprocation between one to the other between positive and negative, 9, through its dance of balancing the interchanging polarity of 3 and 6 in the 3− 3+ 9+ 6+ 6− 9− polar circuit, becomes polarized, thus taking 9 and creating 18 with its polar opposite.

The Enneagram and the Rodin 147 Shears

In the following diagrams, the 18-digit sequences of the Rodin 147 Shears reveal a dance of numerical balance and harmony, containing all sequences, patterns, geometry, and family number groups. The geometry didn't seem to line up with Gator's sequences. The 3's, 6's and 9's only line up with Rodin's 18-digit sequences if seen as two counter-rotating enneagrams (which you will see in the following diagrams). This is similar to the diagrams at the end of Chapter 3 showing the Rodin 147 Shears of 18 digits, minus the 6 digits of the 3 3 9 6 6 9 circuit. These contain the whole 18-digit sequence and show me that the 3 3 9 6 6 9 circuit must be required in order to complete the 1 2 4 8 7 5 geometry missing in the other diagrams, seen as dotted lines. Do you remember me mentioning that I did not know why the two counter-rotating 1 2 4 8 7 5 circuits were missing from the geometry in the diagram at the end of Chapter 3? I still do not really know why, but this may tell us that the 3 3 9 6 6 9 circuit not being present in the geometry may also indicate its incompleteness. We were able to see congruency and completeness in the 12-digits of the 24-digits Phi sequence that did not contain any 3's, 6's or 9's, but that was Phi. It is just an observation.

Here is the first diagram. It is 1 7 3 5 5 3 7 1 9 8 2 6 4 4 6 2 8 9. Within it we can see 1 2 3 4 5 6 7 8 9 creating the "star" geometry in two opposing sequences.

The Enneagram and The Rodin 147 Shears

2 Sequential Circuits
Counter Rotating Apart
From Each Other

Sequence 1: 1 7 3 5 5 3 7 1 9 8 2 6 4 4 6 2 8 9

Extension Emerges
From Sums of 3

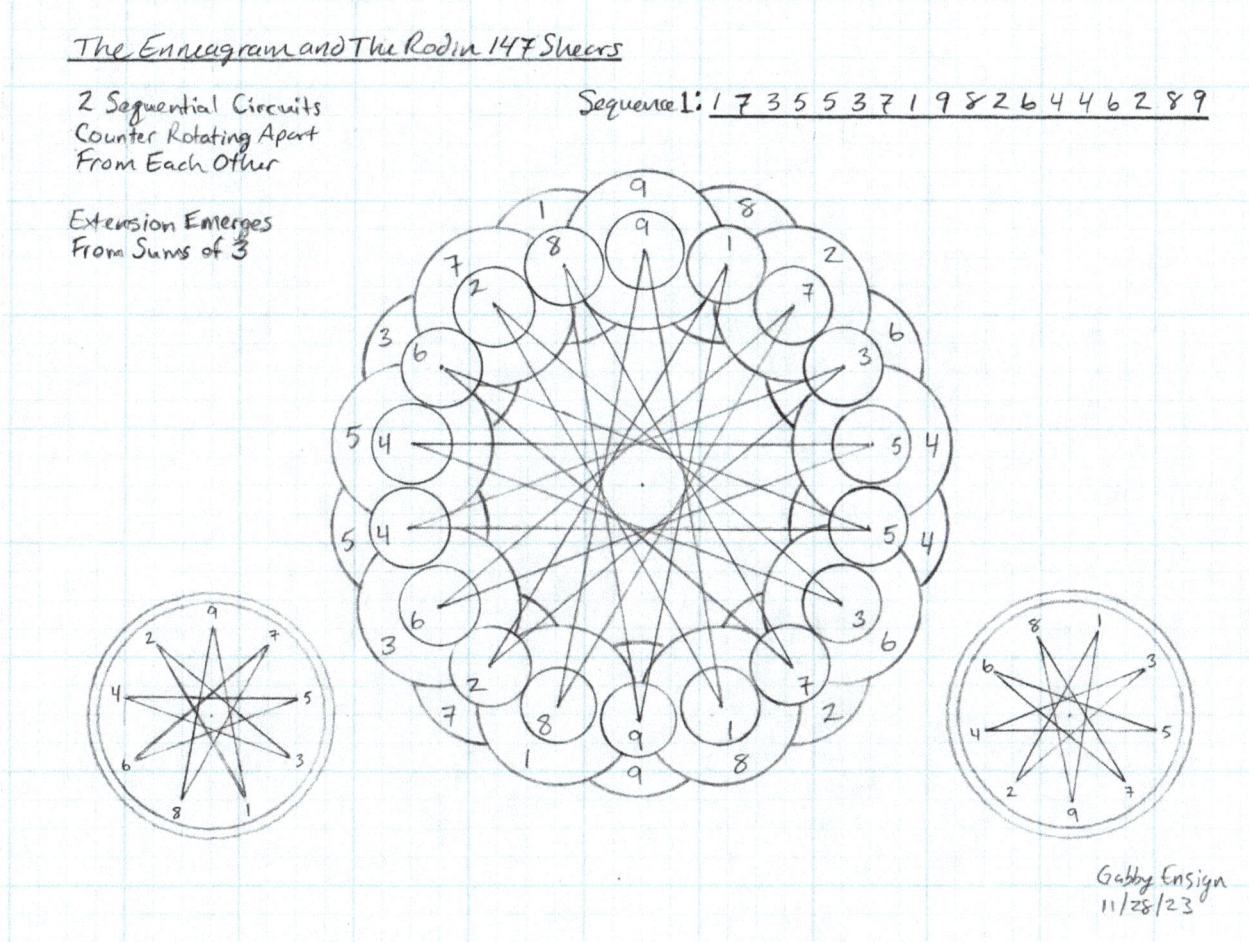

Gabby Ensign
11/28/23

If we look a little closer at one of the two smaller circles at the bottom, the 1 3 5 7 9 2 4 6 8 sequence runs from one point to the next nearest point around the circle, and the 1 5 9 4 8 3 7 2 6 sequence runs every other point around the circle. So within 1 7 3 5 5 3 7 1 9 8 2 6 4 4 6 2 8 9 are the geometry of the polar pairs. So holographic.

The larger circles around the outer part of the diagram are the root sums of any three digits next to each other. This results in duplicating itself as the same sequence directly opposite of itself. The numerical balance of this seen through its geometry is just incredible.

The next diagram is 1 1 6 5 2 9 7 4 3 8 8 3 4 7 9 2 5 6 and follows the same method as the first diagram. This time we see the 1 2 3 4 5 6 7 8 9 sequence moving from one point to every other one around the circle. The sequence 1 3 5 7 9 2 4 6 8 would, in this diagram, create the "star" patterned geometry, and the 1 5 9 4 8 3 7 2 6 sequence is seen going point to point around the circle. Again, only the 1 2 3 4 5 6 7 8 9 sequence is shown, but it is easy to spot the other two in the smaller diagram circles.

The Enneagram and The Rodin 147 Shears

2 Sequencial Circuits
Counter Rotating Apart
From Each Other

Extension Emerges
From Sums of 3

Sequence 2: 1 1 6 5 2 9 7 4 3 8 8 3 4 7 9 2 5 6

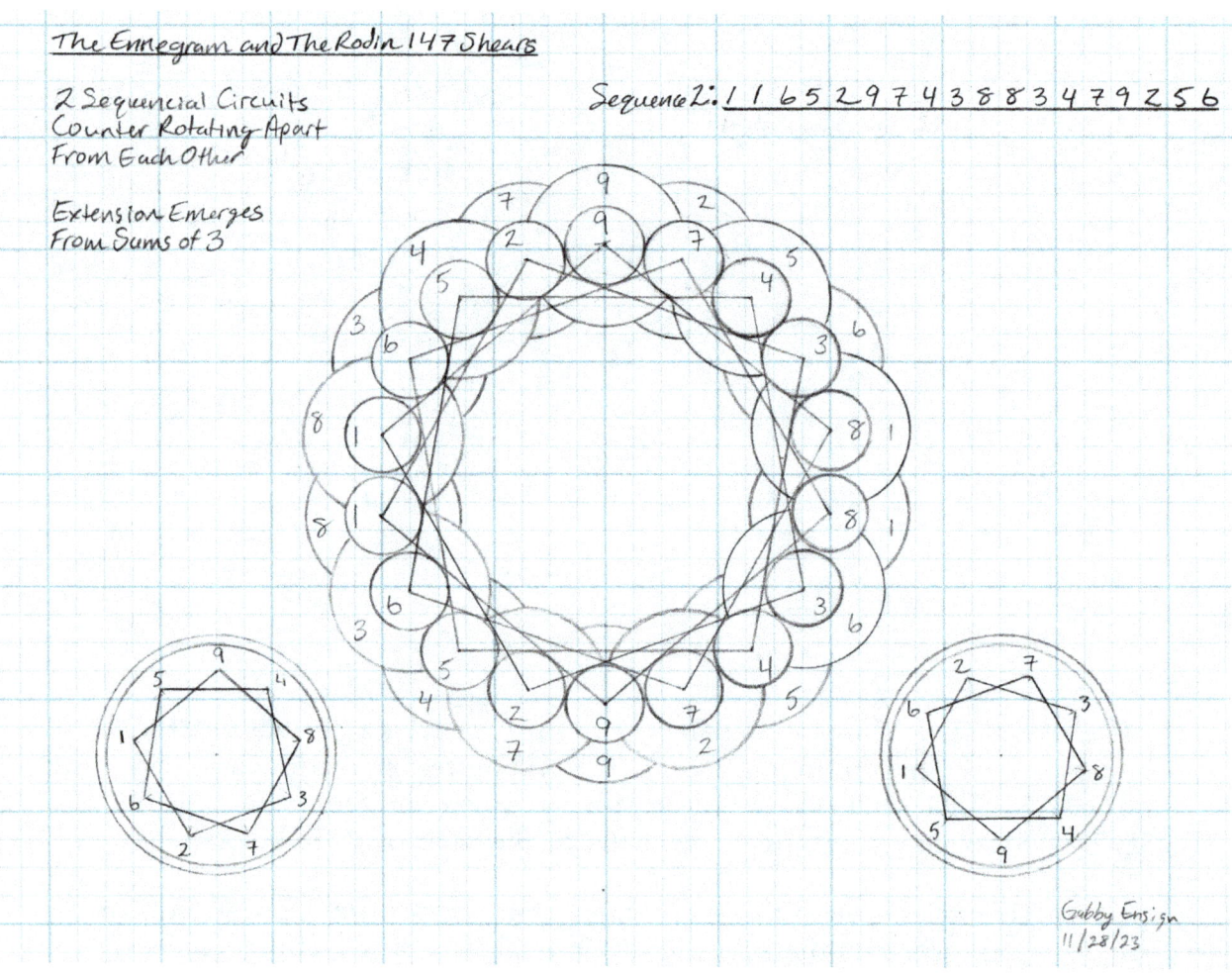

Gabby Ensign
11/28/23

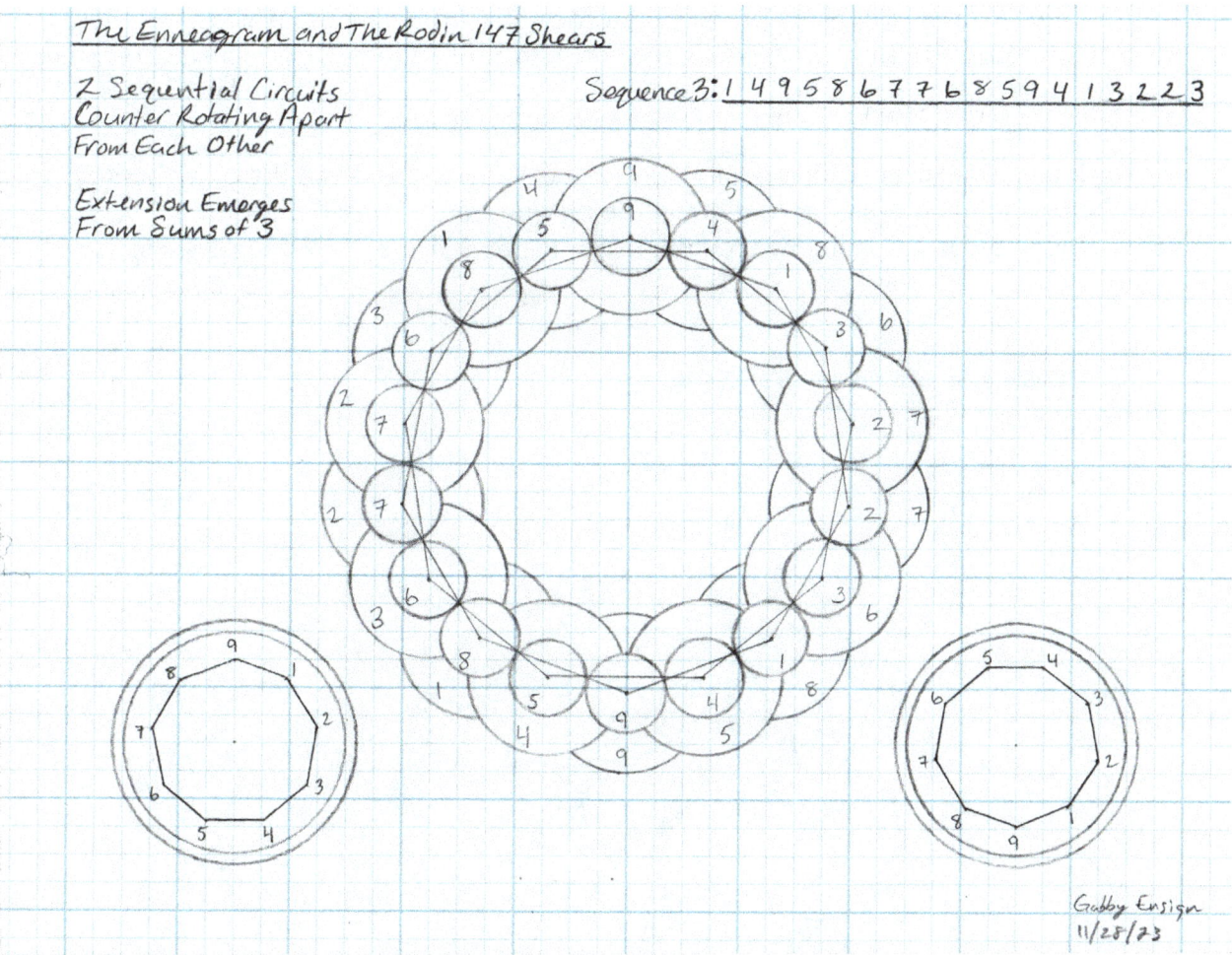

I gave you the third diagram as well because at this point, you should be able to see the rest based on what I gave you with the first two.

Holographic Geometry and the Illusionary Universe

So, all three of the Rodin 147 Shears of 18 digits show the beautiful geometry, and harmonious interconnectedness of the three polar pairs of 9 digits. It is now established and shares the exact same geometry. They are in a sense holographic of each other. Just as the universe is holographic. Holographic in the real sense that it is not real (hey, that was a funny oxymoron, but not really), and also in that by taking any part, it contains the whole. As for nothing being real, nothing in this physical universe is technically real… except YOU! Your experience, your consciousness, your love, and your happiness are real. The rest is just the movement and transfer of energy through time. Everything else, including your ego is not real, because the physical universe and all its dimensions are but a thought form designed to experience separation from your creator. If the ego was created within this universe, of which is not real, then neither is your ego real. Without your ego, what is present underneath is a quiet gentle peace, and a sense of ever-present love. But that topic is for another book. Think of it is a seed planted that may someday bear you fruit.

Back to the next set of diagrams. This gets really cool. I took all three of the Rodin 147 Shears and lined them up to 9 down the middle axis. We get the geometric version of what you saw in the numerical comparison diagram at the end of Chapter 6.

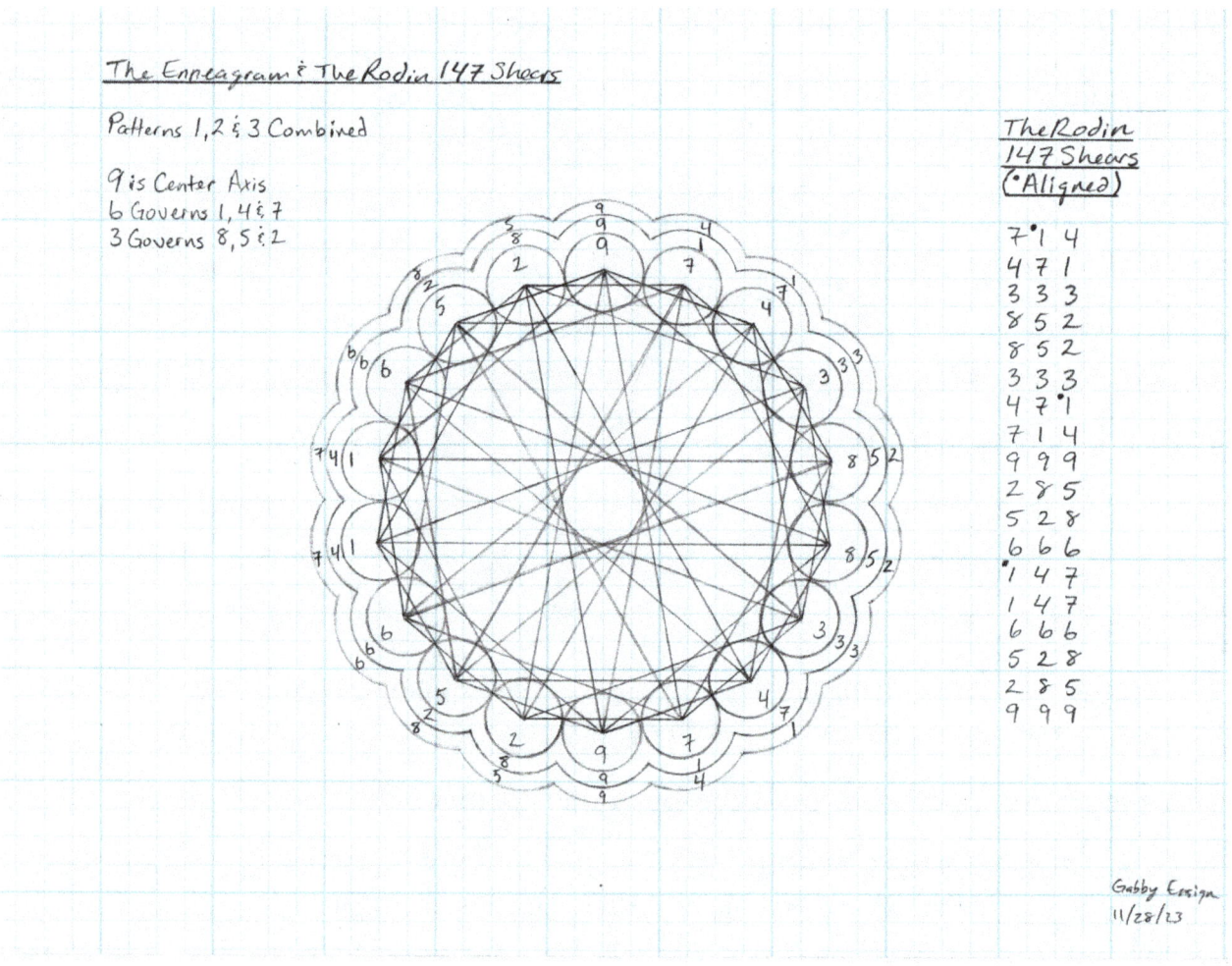

Just as you saw in the numerical diagram at the end of Chapter 6, the Rodin 147 Shears combined contained all three family number groups of 147, 852 and 396. I am a visual person as you can probably tell, and seeing patterns through geometry tells me much more than just seeing numbers. Through geometry, we can better see their relationships.

We can see that 9 is the central axis. We can see that 3 governs 852 and 6 governs 147. We can see the 3 3 9 6 6 9 circuit. We can see the geometries of the three polar pair sequences. What I cannot see is the 1 2 4 8 7 5 doubling circuit. In the numerical diagrams, I had circled 1 2 4 8 7 5 within the sequences and you could see them as every other number. Look again at the 12-digit sequence without the 3 9 6. It is 1 7 5 5 7 1 8 2 4 4 2 8. Do you see it? I will start with 1 and write it this way. 1 8 2 4 4 2 8 1 7 5 5 7. There it is, and the smaller numbers are also 1 2 4 8 7 5 going in reverse as 5 7 8 4 2 1. All three of the Rodin 147 Shears do this, just in a different order of 3. Pretty cool, huh?

The Rodin Trick

I am going to do one more diagram and throw in a wildcard to see if the Rodin symbol fits in anywhere in this method of the Rodin 147 Shears, starting with lining up the 9 at the top axis. You already know it does, don't you?

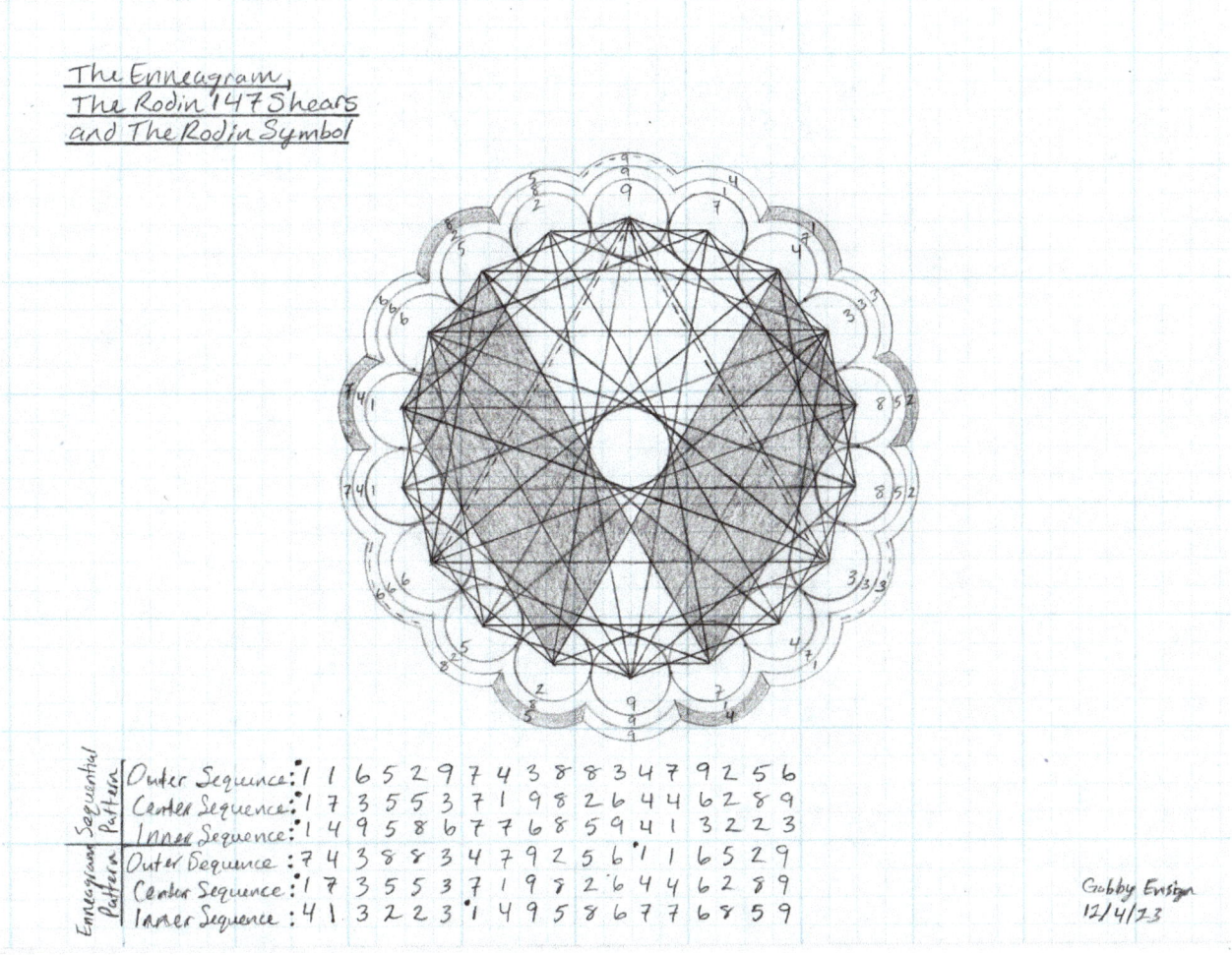

Checkmate. The number 9 contains all of the core sequences in vortex based mathematics through a system of 9 digits. It is also the central point of balance, Source and its numerical form within this universe. If 9 is balance, then through splitting from 9 is the split duality of 3 and 6, becoming separate from 9. The first step in the experience of separation. The division of 9 creates the bipolar fluctuations of 3 and 6, and 9 is the only thing keeping them in balance. And although split and in opposition to each other, 3 and 6 are always kept in balance with each other. In other words, even in a universe where we all think everything is out of balance, nothing is out of balance, except ourselves unless we are brought back into balance by 9 or Source. With our split minds apart from Creation, in the physical holographic universe we call life, it is up to us to tune in to 9 or Creation for balance, or ride the roller coaster of our physical minds of monkey chatter going on in our dualistic split brains of our physical bodies we temporarily inhabit. So again, 9 is stillness, the center, the calm eye of the storm. It is peace.

The Rodin 147 Shears and the Trinity of Creation

I cannot show you how to find peace, but 9 can, if you ask and quietly listen. Well, don't ask the number. You know what I mean. But I can show you what balance looks like in geometric terms within the functions of the 3 3 9 6 6 9 circuit within the Rodin 147 Shears. This brings us back to the Trinity of Creation diagram.

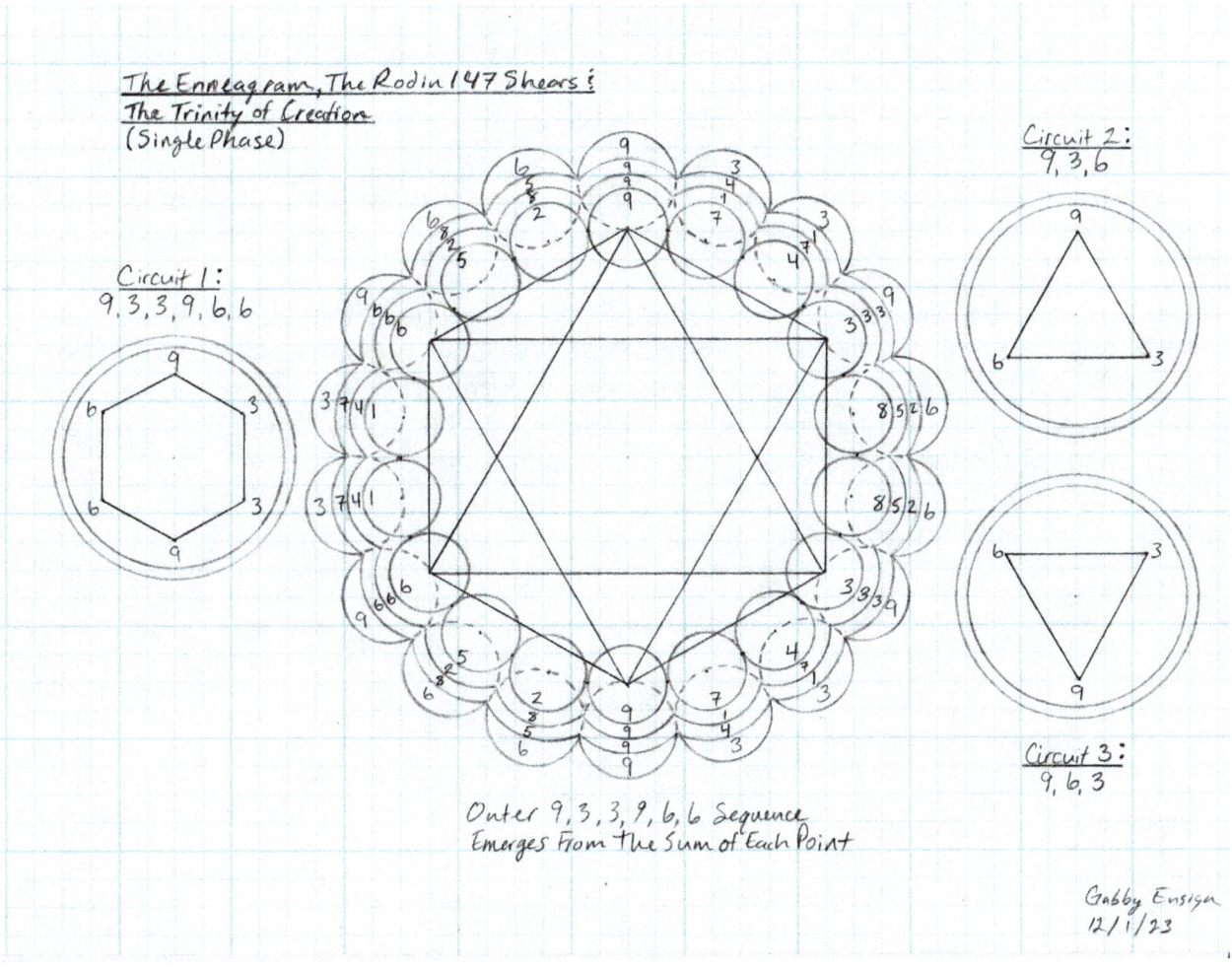

The hexagon that finds itself in so many aspects of nature's blueprint is beautifully laid out within the Rodin 147 Shears. I showed you the 3 3 9 6 6 9 circuit within the sequences and by now it should be easy to spot its triangle geometries. Something interesting happens though if we take the root sum of each of the three numbers aligned within the sequence. Starting at the top with 9 + 9 + 9 = 27 = 2 + 7 = 9, 7 + 1 + 4 = 12 = 1 + 2 = 3, 4 + 7 + 1 = 12 = 1 + 2 = 3, and so on, the 3 3 9 6 6 9 circuit embodies the family number groups within the root sum of their numbers.

The next diagram shows this and also shifts the geometry of the 3 3 9 6 6 9 circuit and its surrounding hexagon to the other numerical points to complete the circle. Circuits 1, 2 and 3 are shown on the left. The second hexagon makes the 5 7 8 4 2 1 reversal of the third hexagon of 1 2 4 8 7 5. The triangles are made up of the other family number groups of 147 and 852. What a beautiful balance of numbers and geometry.

The complete geometry of the Rodin 147 Shears and the Trinity of Creation is seen in the last diagram. If we look between the center hole and the outer points of the toroid, we can see another ring inside the toroid. Not surprisingly, if we measure the distances between the outer points, the middle ring and the center circle, it is a perfect Phi ratio of 1 to 0.618. Lovely. Another harmonic sunflower.

The Enneagram, The Rodin 147 Shears and The Trinity of Creation
(3 Phase Rotation)

Circuit 1:
(1) 9, 3, 3, 9, 6, 6
(2) 7, 8, 4, 2, 1, 5
(3) 4, 8, 7, 5, 1, 2

Circuit 2:
(1) 9, 3, 6
(2) 7, 4, 1
(3) 4, 7, 1

Circuit 3:
(1) 9, 6, 3
(2) 2, 5, 8
(3) 5, 2, 8

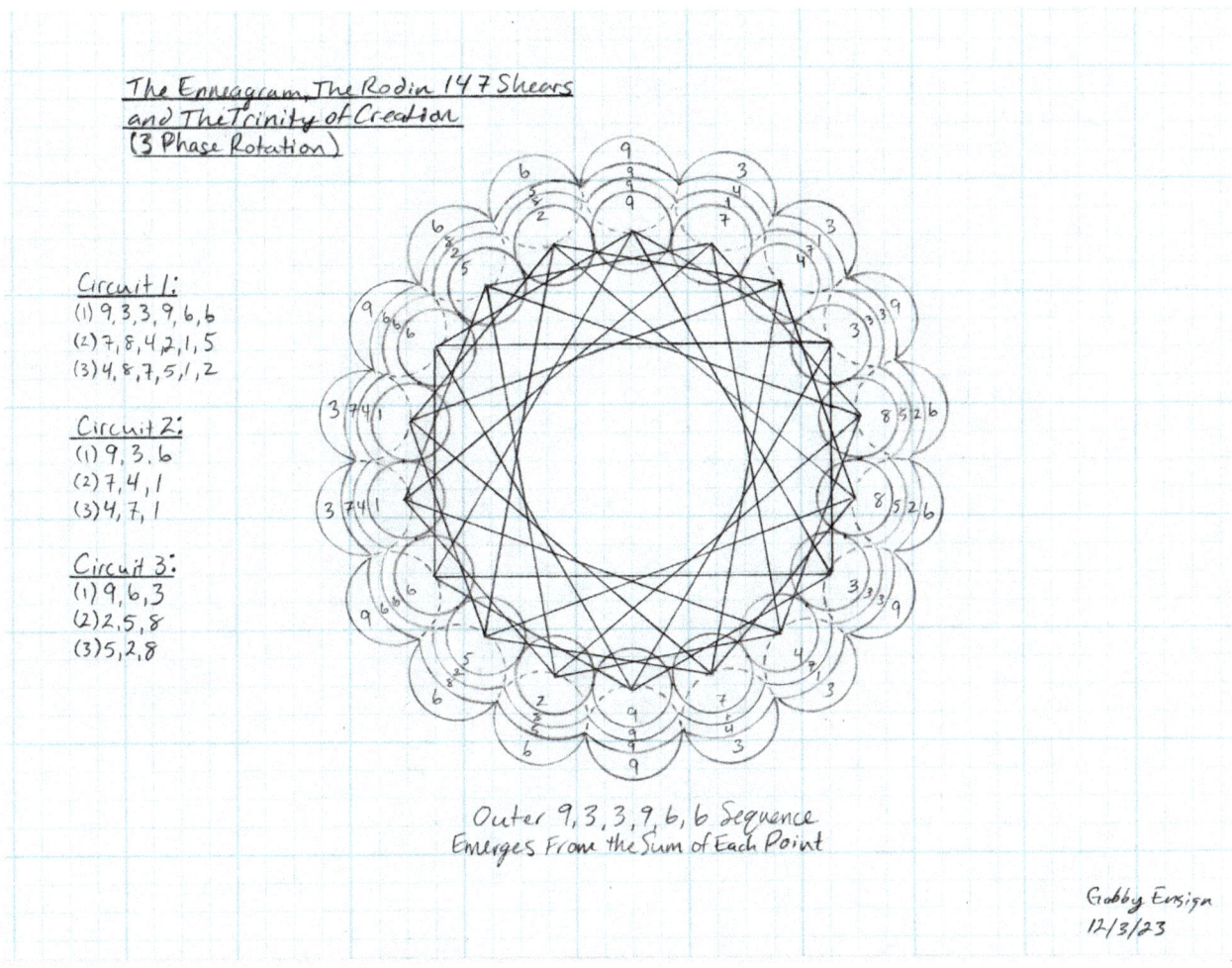

Outer 9, 3, 3, 9, 6, 6 Sequence
Emerges From the Sum of Each Point

Gabby Ensign
12/3/23

The Enneagram, The Rodin 147 Shears and The Trinity of Creation Combined

Outer 9, 3, 3, 9, 6, 6 Sequence
Emerges From The Sum of Each Point

Gabby Ensign
12/2/23

9 Equals Balance, 9 Equals Love

Numbers are interwoven in all things within the universe. The system of 9 digits, with 0 being a representative of 9 because $9 + 0 = 9$, the only other number that does not equal itself when added with 9 is evident within the numbers themselves. The Phi spiral even resembles the number 9 in its curve, like a nautilus shell seen from its side, and is another one of those intrinsic functions that can be seen in just about anything if you were to look in the right place. Other examples of Phi are flat staring you in the face, literally. Your facial structure within the spacing of your eyes, nose, mouth, and chin all conform to Phi in its ratios, and balance is harmony and harmony is love.

Patrick Flanagan, one of the most brilliant scientists and inventors in modern time, said, "Phi is the frequency of love." It has stuck with me ever since I heard him say it, and I believe he is correct.

The last two diagrams of this chapter are as follows. The first takes the 1 through 9 sequence and doubles it, then takes that and doubles it again, except as two separate circuits of numbers which rotate in opposite directions, and rotating opposite to what Gator used in his first inner set of numbers. It seems that no matter which position I put the separate doubled circuit, if counter-rotating, it results in an infinite expansion of 1 2 4 8 7 5 within a toroidal number map. Strange but fascinating. The second diagram is an infinite Phi (Fibonacci) loop.

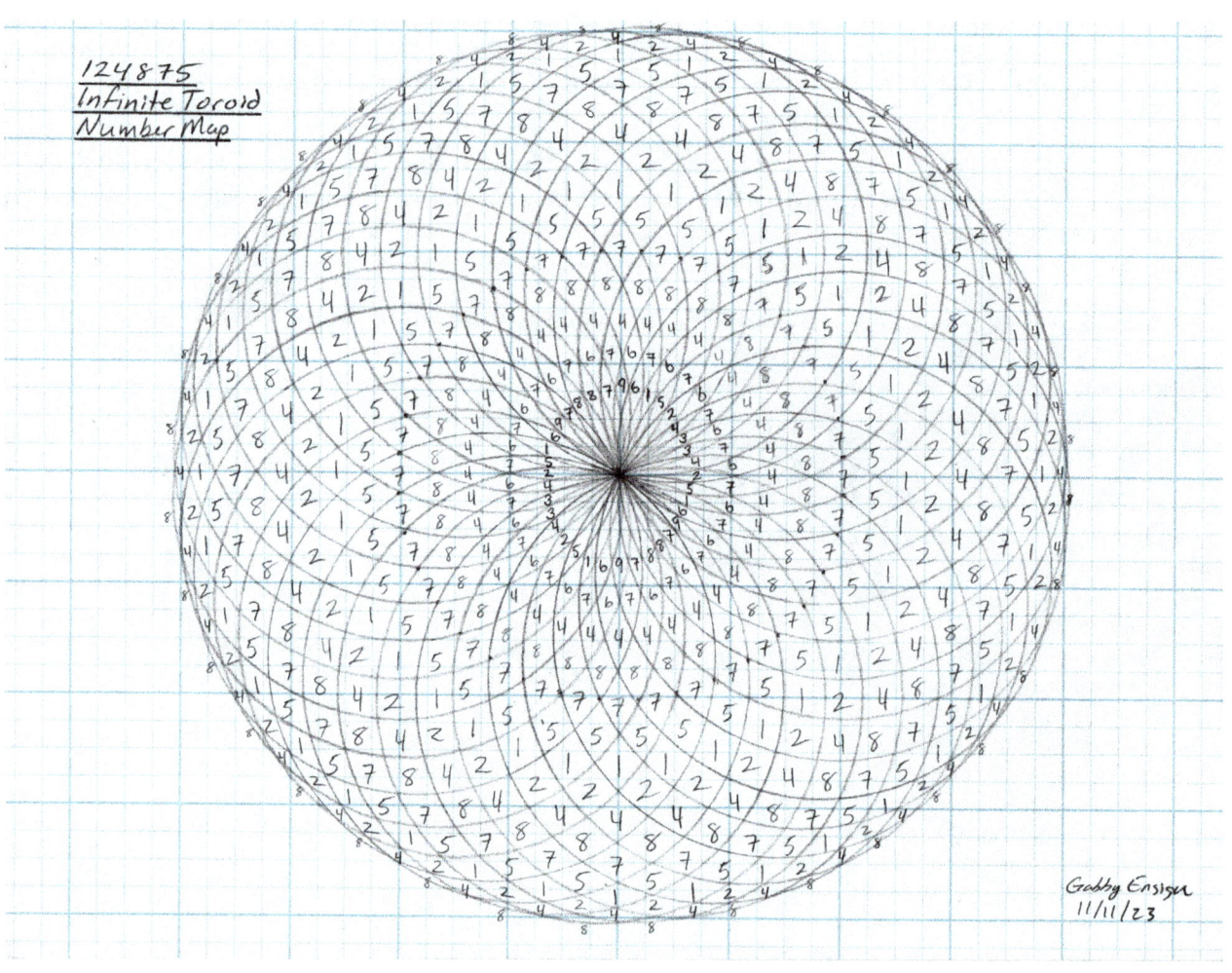

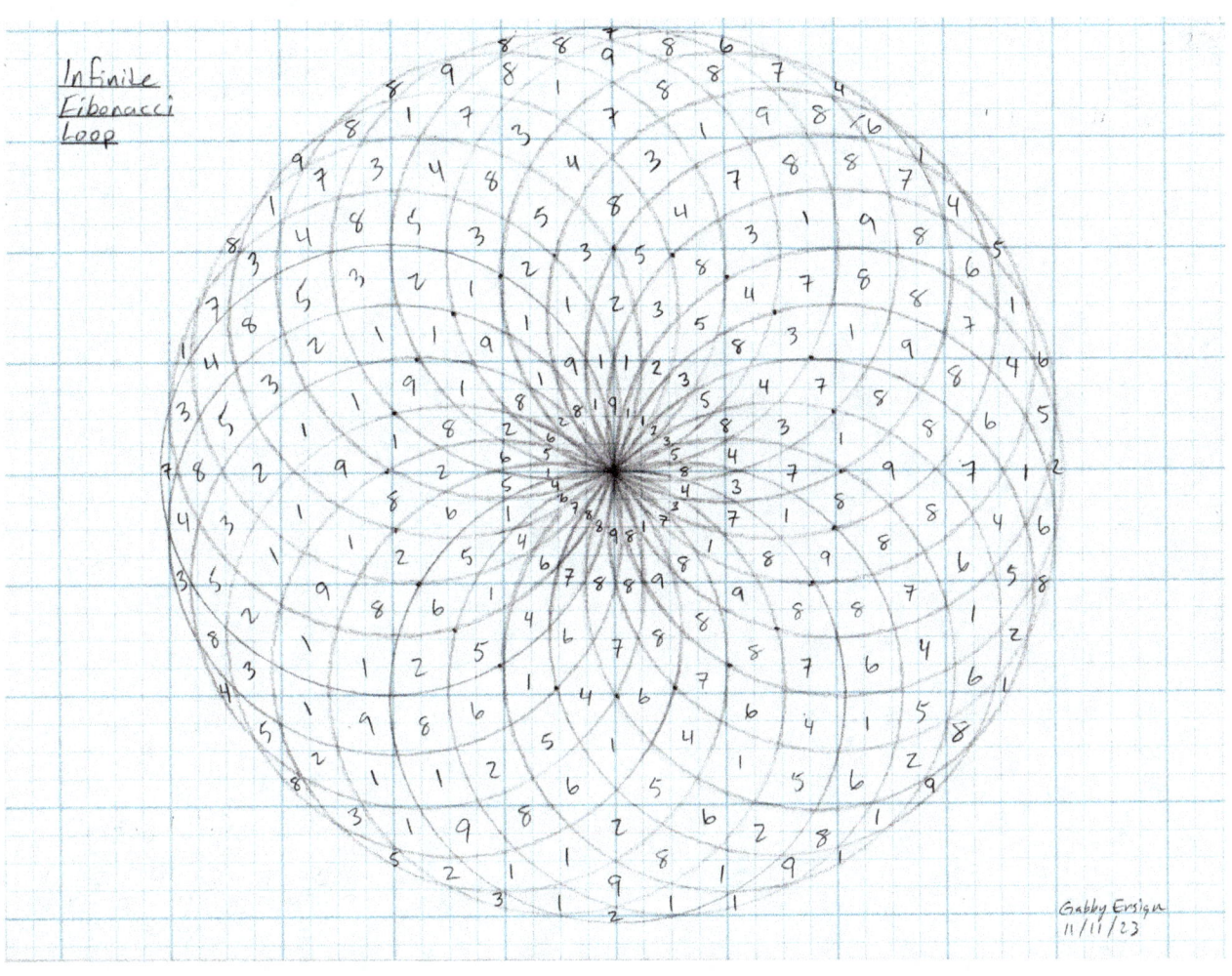

Chapter 8

Vortex Based Mathematics and What Comes Next

Numerical Synchronicities

I am writing this on the 116th day of the year and it is Thursday, April 25, 2024. I learned about vortex based mathematics on August 11, 2023. That was 258 days ago and it will be another 108 days until I reach that anniversary date of one year. In that short time, I had a lifetime worth of knowledge being explained in relationships I had never heard of. With all my previous knowledge, I was able to see and understand it right away, which allowed for faster learning, discoveries of my own, and the writing of this book. It is funny how time works, especially if it is being blanketed in numerical synchronicity like today's date. If today is the 25th, then yesterday numerically was 4-24-24, and there couldn't be a better set of numbers I could ask for. Put that in your back pocket in your mind, and I'll explain what I mean in just a bit. But for now, let us talk real numbers, not synchronicity or hypothetical coincidences.

We have established that the numbers 1 through 9 work together to create the fabric of geometry and mathematics themselves and is evident within the framework of vortex based mathematics. We know that those 9 numbers can be arranged and organized into more complex patterns, both geometrically and numerically, to create fundamental structures from Phi and it's 24-digit harmonics. I have explained many relationships of all of the sequences that make up the Rodin 147 Shears and the polar pairs. I have even established the relationships behind the harmonic numbers chart in music and hertz, to the movement and distance of celestial objects in our solar system, to a few fundamental math principles like the speed of light.

Even the ancients built structures in a geomathematical code system using many of the same numerical concepts and harmonic ratios and numbers. All this is reflected in the world we live in. In this final chapter we will explore more real world examples and breakthroughs in science which will pave the way for a future working with nature using conscious technologies. Let us relearn what our ancient ancestors knew, and with the technology we have now, we have a chance to change our future.

The Ferrocell

The lines of force within magnetism show us another view of the toroidal geometry behind the movement of energy making up the Archimedean solid like structures in the universe. A circle or curve is the feminine aspect, whereas lines and angles are the masculine aspect. Combined, they form another part of that balanced universe we were talking about. Cymatics is a beautiful measure of both in its creations through resonant sound frequencies in a medium like sand or water. Another is the ferrocell, which gives us the ability to actually see these lines of force within magnetism, instead of merely observing its effects or seeing a computer model. This has never happened before within a real time visual display. Below are neodymium ball magnets glued to the end of a wooden dowel and placed on top of a ferrocell. Light is reflecting off a liquid ferrofluid between two plates of glass, showing the magnetic field of the magnet.

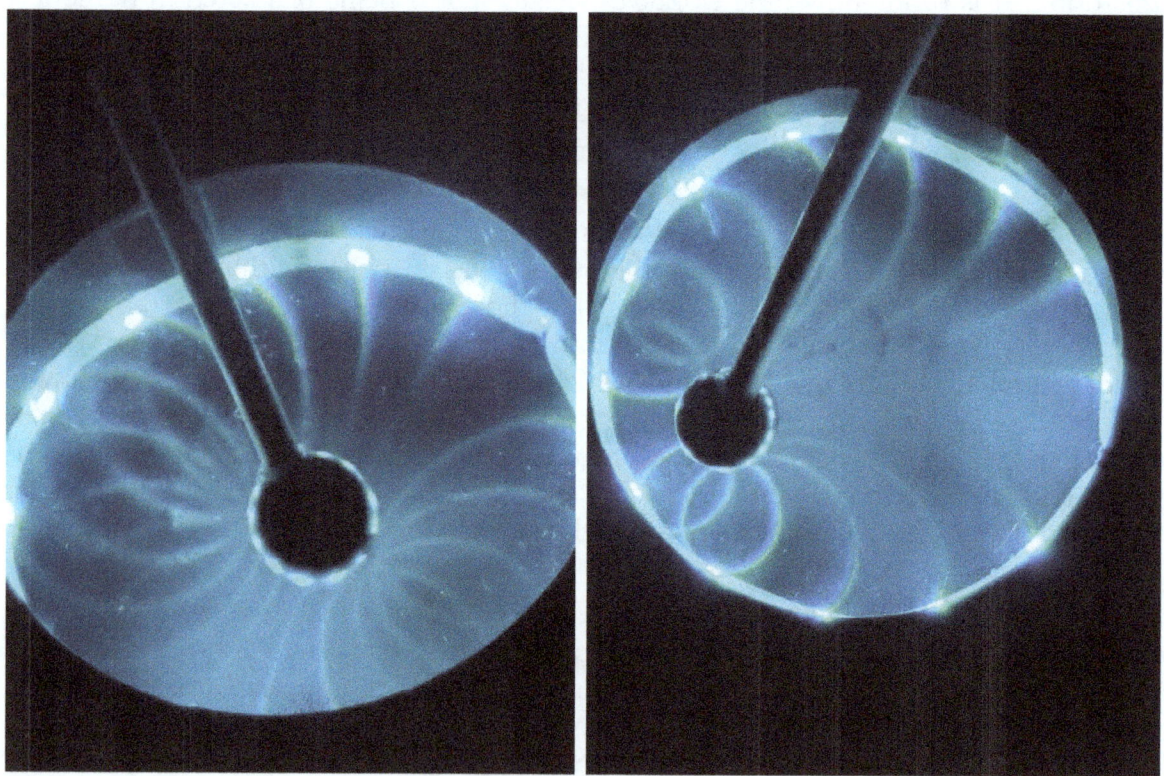

Image 40 – 41: Magnetic Fields Seen Through Ferrocells

Seen with your own eyes, this is more incredible than any of the videos you will see online. I built a ferrocell display and a viewing box with my friend John, and what I was seeing blew me away. The three-dimensional effect gives you a full 3-D view of a toroidal magnetic field.

Image 42: My Ferrocell Display Box

The Hopf Fibration

The toroid shape is also a fundamental shape in the universe according to the famous physicist, Sir Roger Penrose, Noble Laureate in physics. He explained something called the Hopf fibration.

Image 43: The Hopf Fibration

The Hopf fibration was discovered by Heinz Hopf in 1931. Made up of Hopf fiber bundles, they are evident in at least 8 different scenarios within physics. They are found in qubits within the two level quantum systems, the Dirac equation, Wignerism through helicity representations, gauge symmetry, magnetic monopolies, twister theory in Robinson congruences and Penrose mechanics within a harmonic oscillator, and even general relatively in the Taub-NUT space. Sir Roger Penrose described the Hopf fibration as, "An element of the architecture of our world." [27]

Within the feminine aspect of Phi seen in its curve and toroidal structure, we get an equal balance of the male aspect of 1.618 in ratio and numerical terms, along with its straight line geometries. The one other geometry that can hold and explain all other geometries is known as the cuboctahedron, or vector equilibrium.

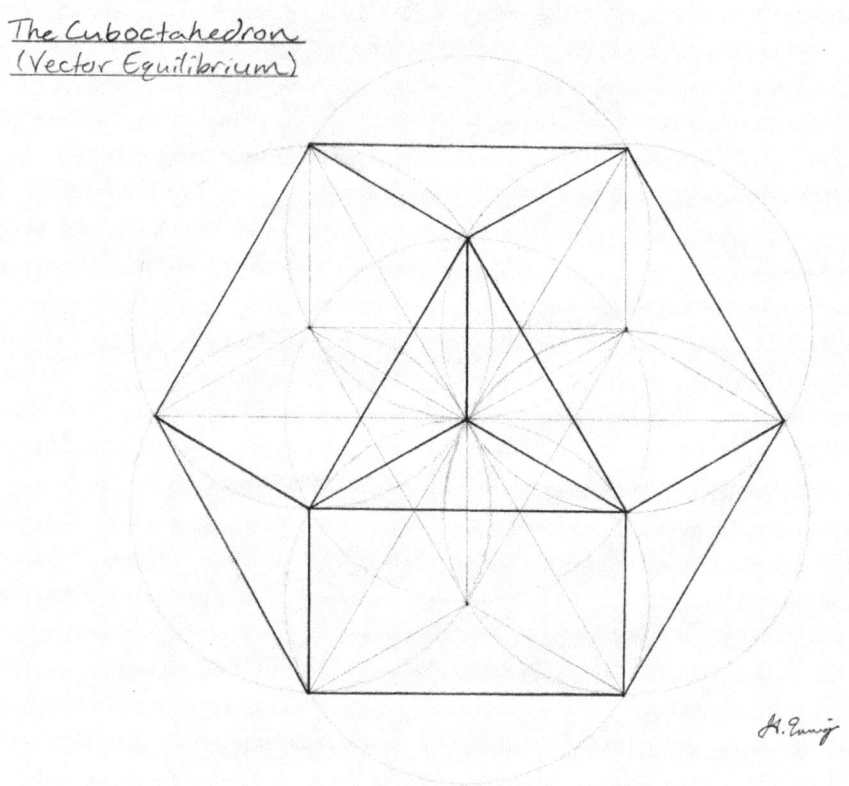

Image 44: Cuboctahedron

Buckminster Fuller and Sphere Packing

Buckminster Fuller, a famous mathematician, inventor and the creator of the geodesic dome like the one seen at Epcot Center in Disney World, discovered that the cuboctahedron is the only polyhedron in which the distance between its center to the vertex is the same as the distance between its edges. If we were to look at the vector equilibrium, it has the same length vectors in three-dimensional space.[28] It is this shape that functions in tandem to the energetic structure and flow of the toroid. The Buckminster Fuller Institute describes something called packing, in which equal radius spheres can fill these polyhedra in three dimensions.

"Omnidirectional concentric closest packing of equal radius spheres about a nuclear sphere forms a matrix of vector equilibria of progressively higher frequencies."
~ Buckminster Fuller [29]

We see this effect quite easily in magnetism if we were to place a spherical magnet on a suspended plane and each magnet placed within the plane packs closer together into tightly arranged, but evenly spaced geometries.[30]

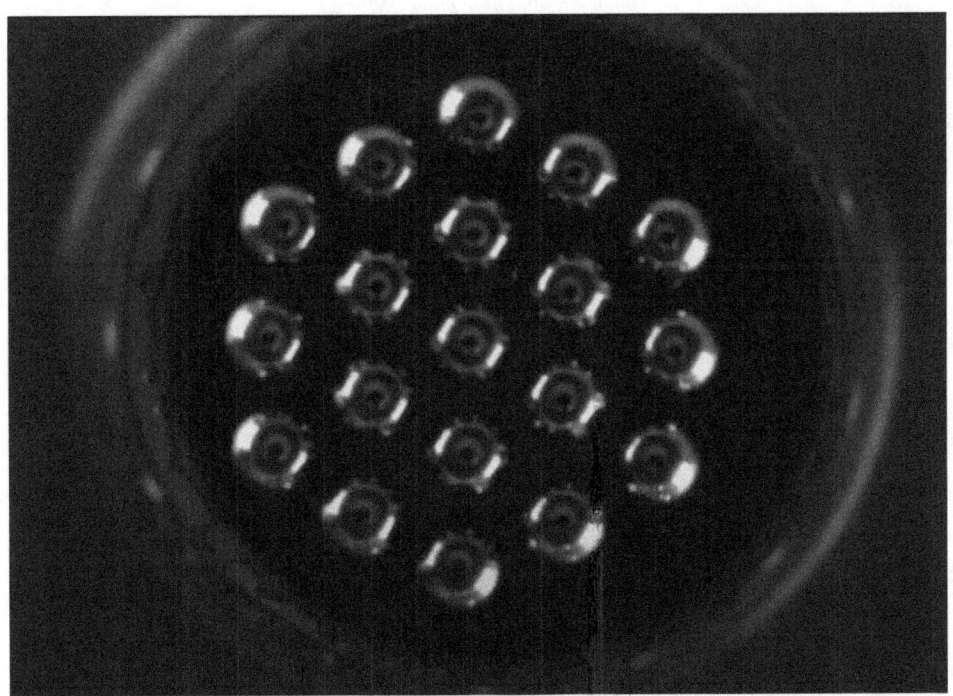

Image 45: Magnetic Packing

The force of magnetism creates the optimal conditions for the most efficient placement of "particle" matter. To nail this point even further, we can look at the hexagonal shape and structure of nanographene molecules through an atomic force microscopic image, discovered by scientists at IBM.[31]

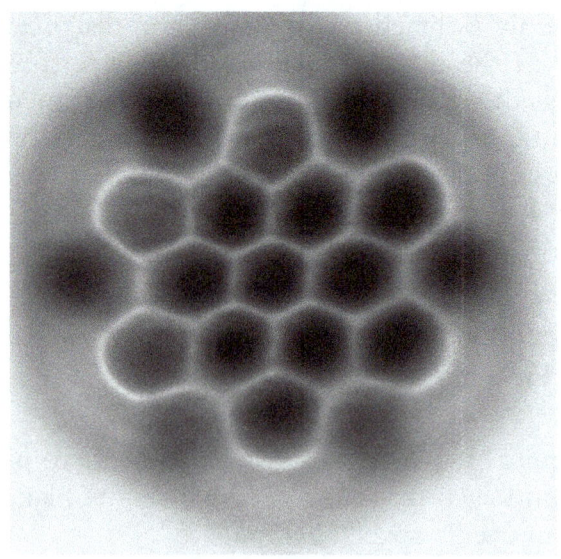

Image 46: Nanographene Molecules

It is the exact arrangement of the natural placement of magnetism within the 19 sphere arrangement (18 + 1 central sphere) seen in the previous image. We can see even further evidence in the packing of atoms in the research done with electron microscopes.[32]

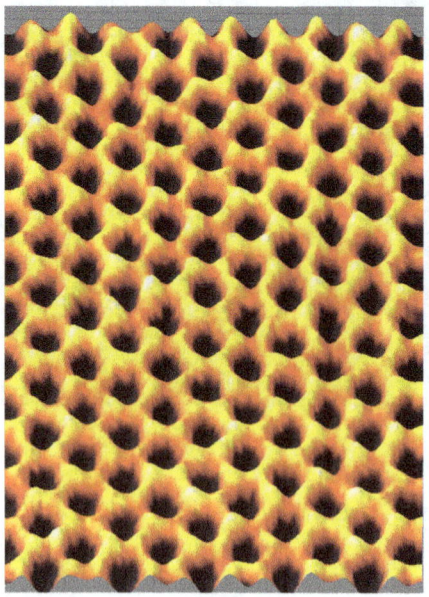

Image 47: Electron Microscopy of Graphene Lattice

In a video by Casey House of the Design Science Studio at the Buckminster Fuller Institute, entitled, "The Synergetic Geometry of Buckminster Fuller/Intro," he shows the results of packing done with the fundamental Platonic polyhedrons, and the geometry formed by connecting the "kissing points," or the points in which the spheres touch.[33] The following image I drew shows these relationships and included is the cuboctahedron.

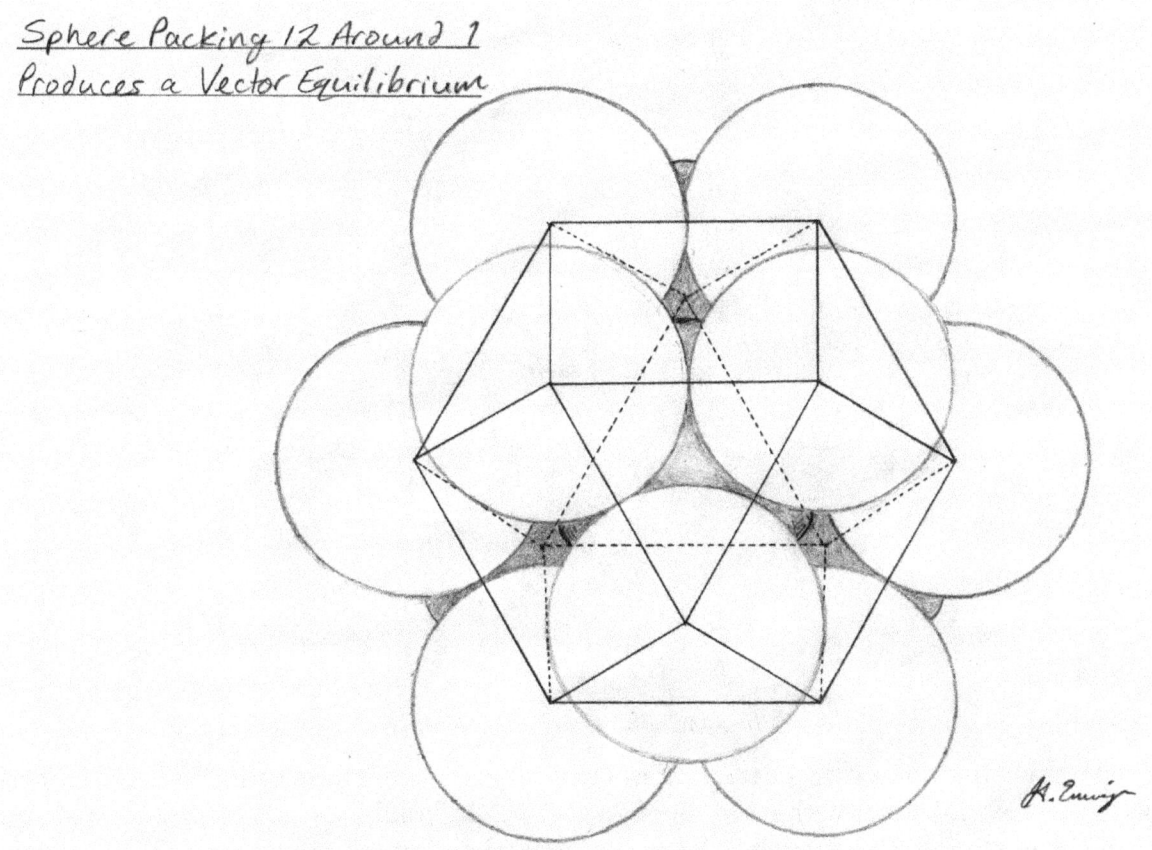

Image 48: Sphere Packing of Cuboctahedron

Remember, at the beginning of the book I showed the first stages of the creation of a salamander through mitosis and the doubling circuit? It is using the same forces, math and geometry. The flower of life is a perfect model of the stages of cellular mitosis.

If you take the first 7 complete circles of the flower of life, which is made of 1 inner circle and 6 outer circles, then add another layer of circles to it and you get 12 outer circles, which is 19 total circles. This is the egg of life in spherical form if seen in three dimensions. This completes a total of 8 spheres or 8 cells in cellular mitosis.

Image 49: Mitosis in the Egg of Life

We can continue along the outside and we will get 18 outer circles with 37 total circles with the next, 24 outer with 61 total, 30 outer with 91 total, 36 outer with 127 total and it continues outward infinitely. Minus the central circuit or sphere, we get harmonic numbers in this arrangement starting with 6, then 12/18, 18/36, 24/60, 30/90, 36/126 and so on. You see, even packing spheres contain harmonic geometry. The lines and points that connect any given circle or sphere create the intrinsic geometries.

"We have a mathematical phenomenon known as a geodesic. A geodesic is the most economical relationship between any two events. A special case geodesic finds that a seemingly straight line is the shortest distance between two points in a plane. Geodesic lines are the shortest surface distances between two points on the outside of a sphere. Spherical great circles are geodesic."
~Buckminster Fuller [34]

There is much to learn from Buckminster Fuller in the next quote.

"All the calculations required for the design of geodesic domes may be derived from the three basic triangles of the three basic structured systems: The 120 right spherical triangles of the icosahedron, the 48 right spherical triangles of the octahedron and the 24 right spherical triangles of the tetrahedron."
~Buckminster Fuller [35]

Nassim Haramein and the 64 Tetrahedron Grid

The importance of the vector equilibrium and the work of Buckminster Fuller was paramount. We see similarities in the work of Nassim Haramein, a physicist and a leading contributor to a working unified field model.

"The 64 tetrahedron grid bloomed into a whole new understanding of the structure of the vacuum. It was exactly what I needed. It had an equilibrium at the center that was actually surrounded by another equilibrium, and you could visualize the structure growing in perfect fractal octaves. From infinitely big to infinitely small. I had not only my polarities producing the equilibrium and singularity, but as it did, it produced fractal structures that could be scaled from infinitely big to infinitely small. I felt I had found something very profound. It was a true three-dimensional fractal structure and it grew in perfect octaves. I could see that this metric could actually be built in a completely different way as well. It could be built out of 8 star tetrahedrons, each made out of 8 tetrahedrons each. The eighth tetrahedron coming together produces the 64 tetrahedron grid. Eight star tetrahedrons are built by 8 tetrahedrons and have their tetrahedron pointing out, radiating out, and when they come together, they produce a vector equilibrium in the middle, which is 8 tetrahedrons pointing in. So, I had now both of the sides of the event horizon that produced a boundary condition. The radiative side and the contractive side of the event horizon. I thought, this is it. I found it and I got very excited. This has all the elements necessary for me to be able to map out and describe the structure of the vacuum. The key to the knowledge necessary to understand creation."
~Nassim Haramein [36]

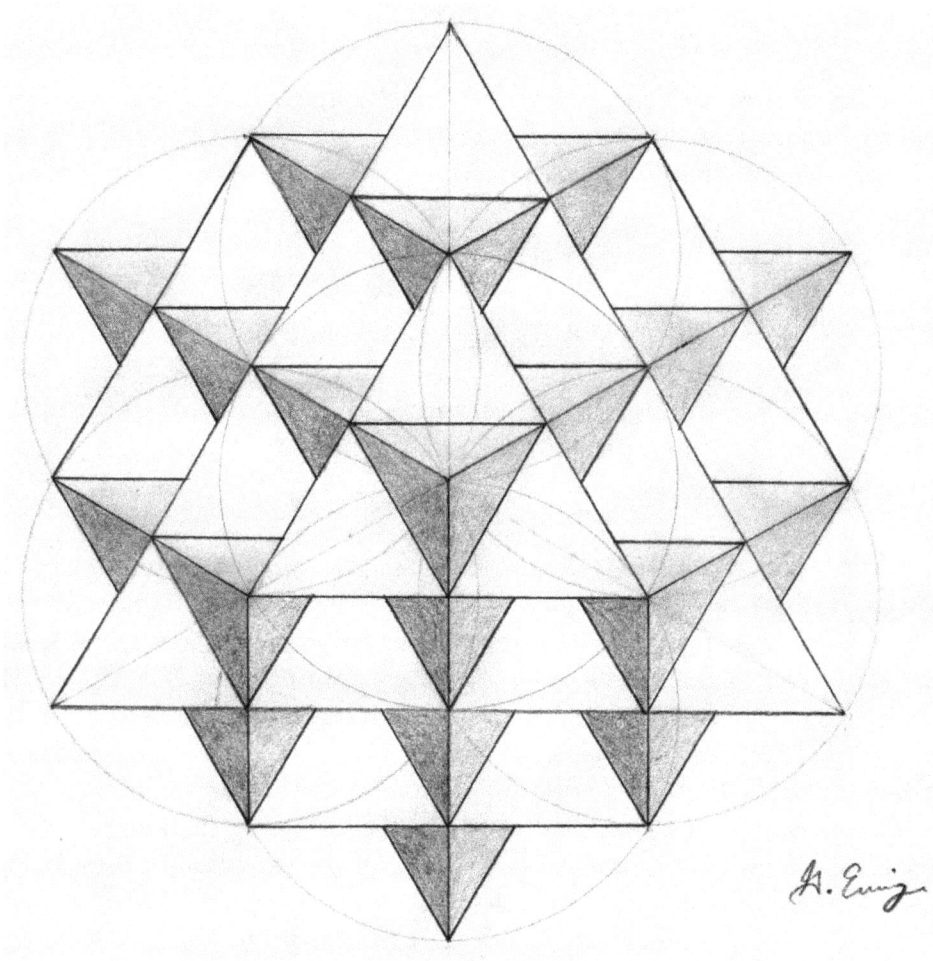

Image 50: The 64 Tetrahedron Grid

You can find more about Haramein's work by visiting the Resonance Science Foundation website at http://resonancescience.org.

The Vector Equilibrium

In vortex based mathematics, the cuboctahedron shows the relationships to the family number groups within its faces by calculating the root sum of its faces surrounding any central given face to equal that central number on a 3 and 6 centrally aligned axis.

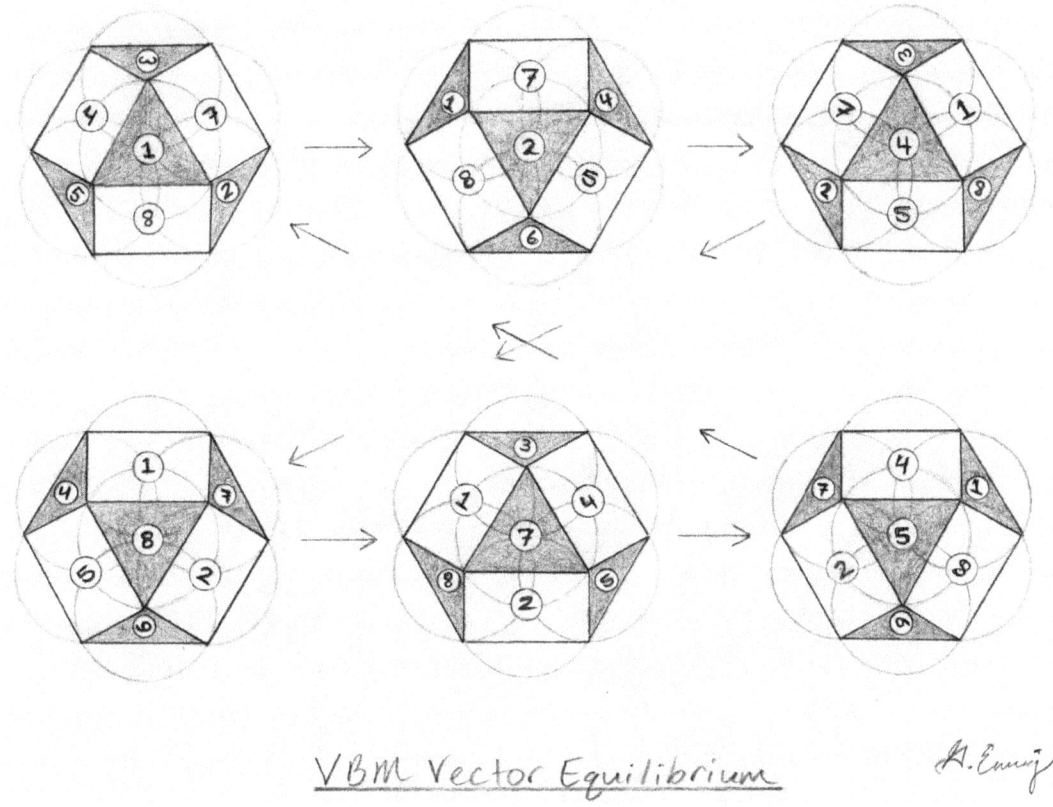

Image 51: Vortex Based Mathematics Vector Equilibrium

Luke Ragle created three-dimensional modeling in an animated graphic entitled *"Vortex Based Mathematics Vector Equilibrium."* [37] We see the pole of 3 with the surrounding faces of 1, 4 and 7, which equal the root sum of 3, 1 + 4 + 7 = 12 = 1 + 2 = 3.

Within the surrounding squares and triangles, it holds the 1 2 4 8 7 5 doubling circuit. The opposite pole of 6 with the surrounding faces of 8, 5 and 2 will equal the root sum of 6, 8 + 5 + 2 = 15 = 1 + 5 = 6. This assignment of numbers on the faces of the vector equilibrium is similar to the diagram I showed in Chapter 4 emulating the first steps of cellular mitosis in the diagram, "Rotating 1 2 4 8 7 5 on a 396 Axis."

Spiraling Outward

Before moving on with the vector equilibrium, I will have to take you back to 9 for a second. Just wait, the grand finale is just around the corner. See if you can spot the two outward spiraling sequences in the next diagram before I try to explain it.

Duel Spiraling Sequential Sequences Showing Spires of Equal Root Sums

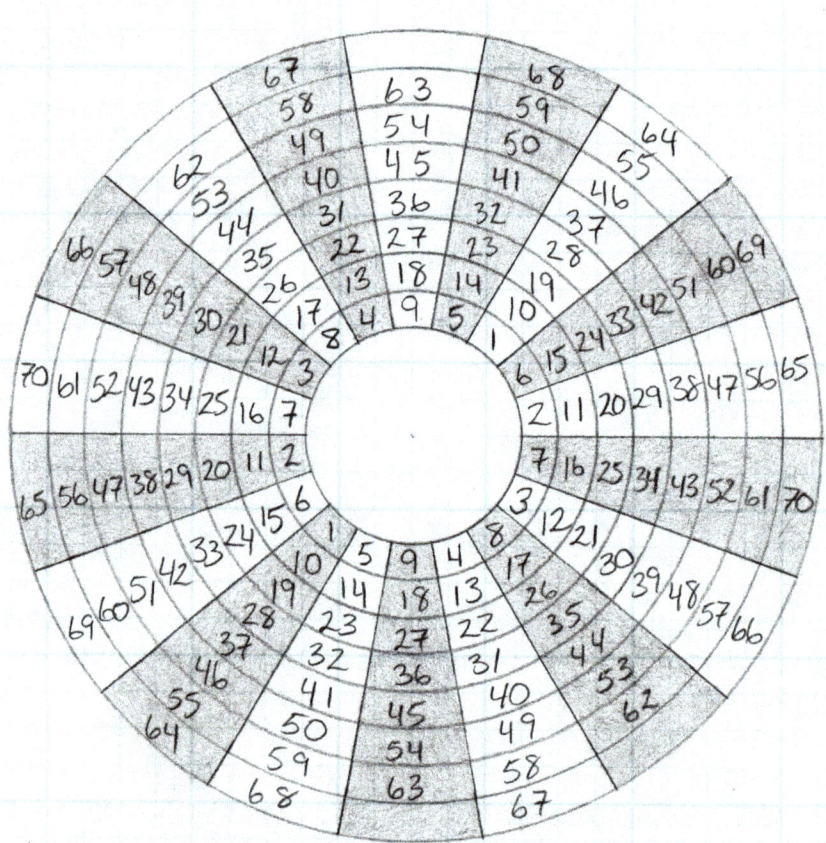

Gabby Ensign

In Chapter 6, the diagram, "360° Circle, The Pentagon and the Polar Pairs Sequences," shows us that within each rotation of the Pentagram and Pentagon sequences around the circle, if continued in its sequence by an "up-step" into a higher dimensional plane of rotation, results in the continuance of the sequence in alignment with its previous rotation around the next outer circle. Within this diagram I made the white tiles move clockwise around the circle, then step up to 10 above the 1 in the first inner circle. The black tiles do the same thing in the opposite direction on its mirror. The sequence 9 5 1 6 2 7 3 8 4 9 5 1 6 2 7 3 8 4 is also the Rodin Symbol copied to its 180° mirror rotation of itself, effectively doubling the 1 through 9 sequence to its 18-digit harmonic. Notice that each spire moving outward from the most inner circle is the same number when calculating its root sum.

Robert Edward Grant and the Prime Wheel

This last diagram, along with the pentagon diagram just mentioned, was expressing a fundamental secret that contained the instructions to understanding all the most harmonic numbers in the universe with the constants of nature, while predicting all primes, in a dance and song of music in geometric form. The harmonics of 24 in a waveform using this method (just like what was seen in the functions of Phi within its 24-digit circuit or sequence) gives us the answer. I was having fun with two sets of 9. It just so happens the vector equilibrium is contained within 24 outer points. Had I done the same with 24 points I would have made the same discovery.

This discovery was made by Robert Edward Grant and his team. He is an author, inventor, mathematician, businessman, and guest speaker on Gaia TV and other programs. Grant models the directional pathways in a cuboctahedron, using a 24-digit sequential pattern and rotation around its central axis as a model. In other words, he wrote 1 through 24 around a circle (which equates to 360° ÷ 15° = 24 points), like the hours on a 24 hour clock, and when he got to 25, he stepped up to the next outer circle above 24 to start the second rotation of the circle in the clock. This would expand out infinitely in theory.

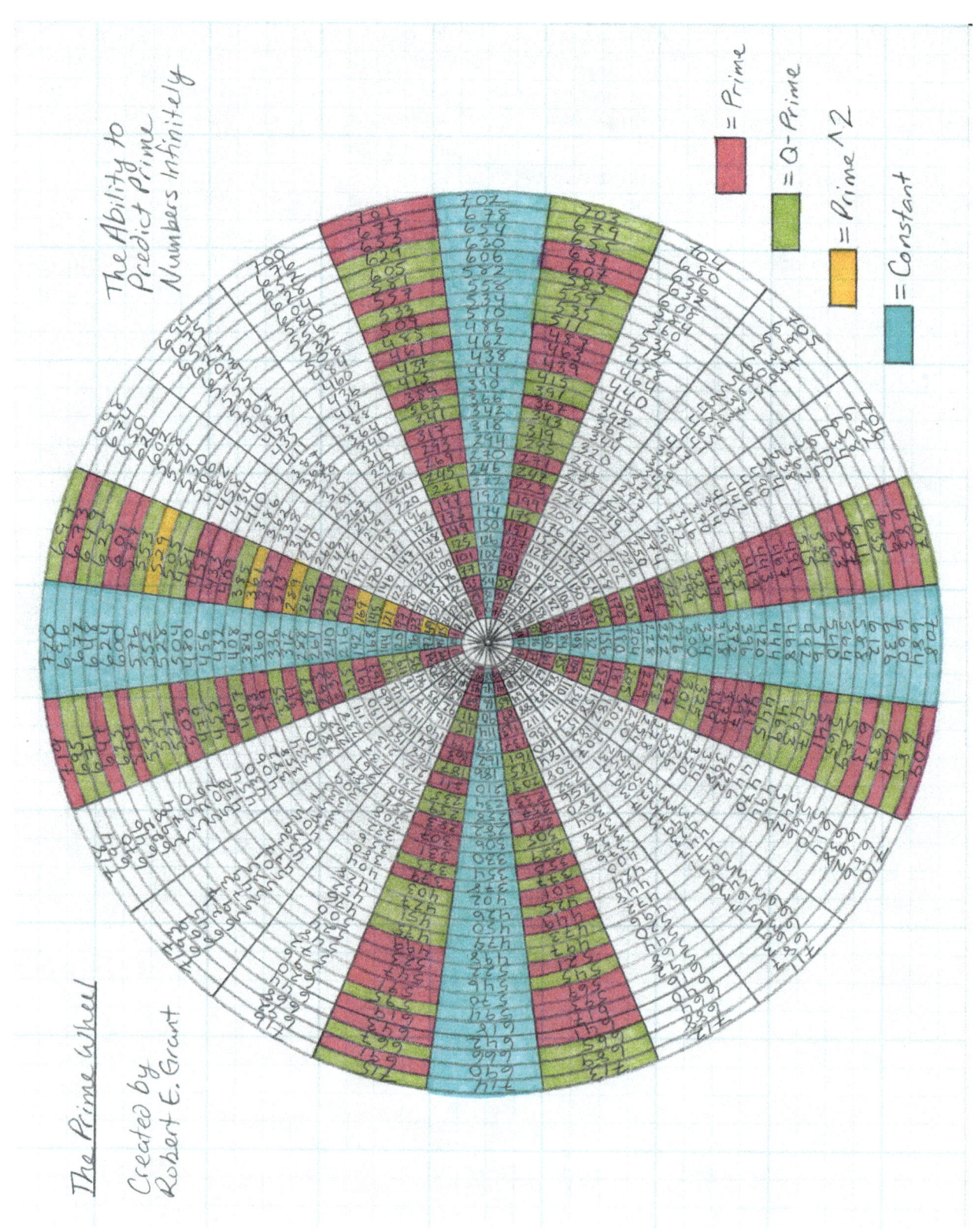

Grant calls this, "The Prime Wheel," and it not only gives us all the known 80 some mathematical constants, it gives us all the most harmonic numbers, prime numbers, and quasi-prime numbers in perfect geometric alignment with each other, and all the musical notes. He adds color to make it obvious where the relationships are in the circular graph.

It was never thought possible, but Grant discovered the ability to predict prime numbers infinitely.[38] In *"Fundamental Constants Mirror Transform Infinitely in Mirror Symmetry,"* he explains the following:

"You start with 137.5°, that's the golden angle. What I'm saying to you is it is almost like looking at a prism. Let's say you start with a prism at one side of it and you see Alpha, you turn it slightly and you see another side of it and guess what that other side is? Phi. Then you look at it this way (rotated) and you see Pi. You (rotate again) look at it this way and you're seeing the Euler Maseroni number. You're seeing all the fundamental math constants and guess what that prism actually is?

"It's the vector equilibrium.

"The vector equilibrium has 24 edges, the flower of life has 24 circles on its perimeter. The 24-hour clock is just a two dimensional form of the vector equilibrium.

"Stephen Hawking said that you can't have a unified physics model unless there is a unified constant model to understand how the constants work with each other. What I'm telling you is that the constants are just different prism reflections of the same number (of one)."
~Robert Edward Grant [39]

Vortex Based Mathematics and the Prime Wheel

I decided to convert his chart, which he derived from his understanding of vortex based mathematics, into their digital root sums.

The Prime Wheel
by Robert Edward Grant
(Digits' Root Sums)

- 24 single digit rows overlapping clockwise around the circle, shown as their digital root sums equalling 1 through 9.

Family Number Groups:
396, 147, 852

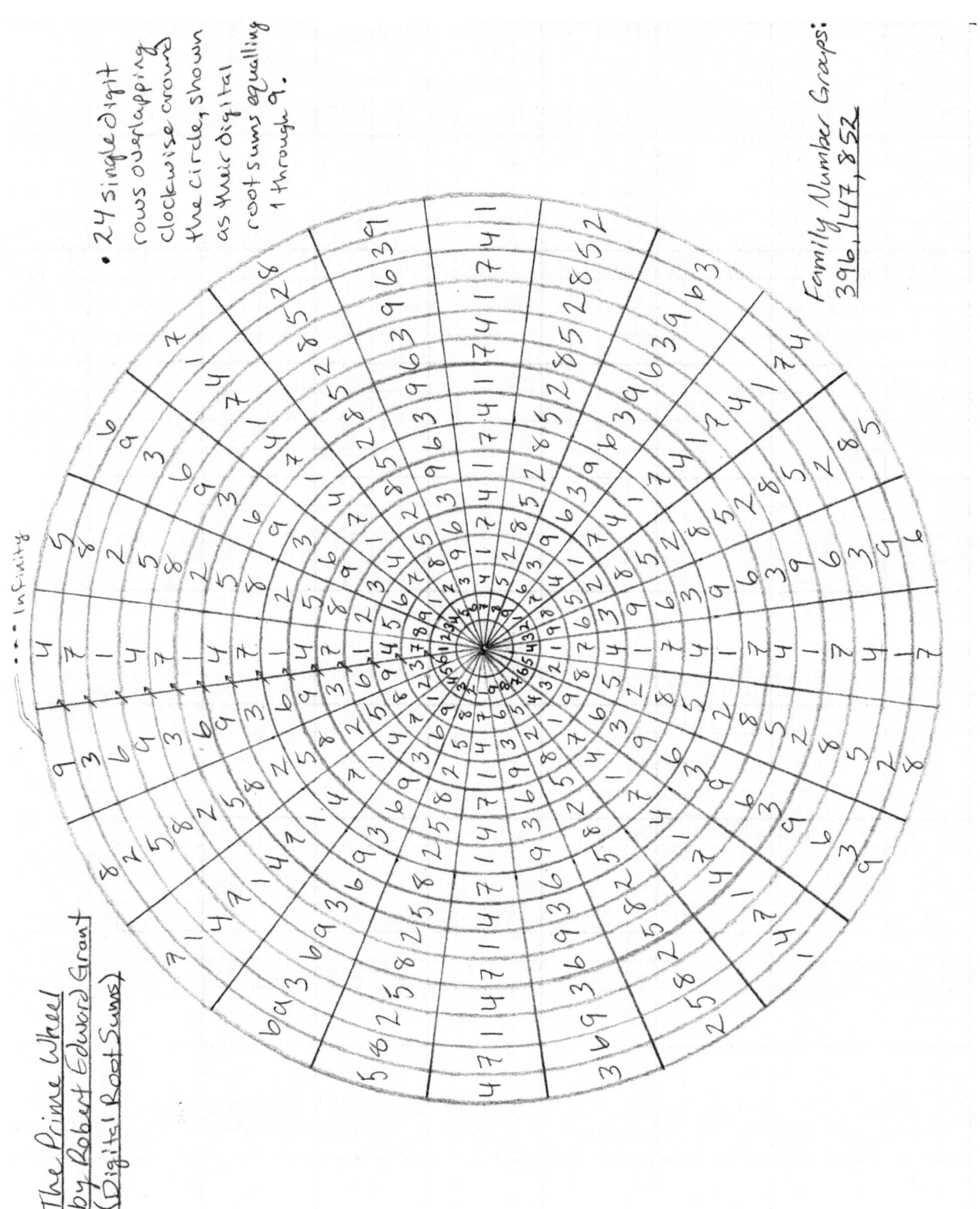

After completing this diagram I discovered that Grant created something very similar depicting the three family number groups. This diagram of the root sum of the prime wheel helped me further understand the relationship that vortex based mathematics has with the fundamental fabric of the universe. The Prime Wheel solves so many missing pieces of the puzzle, especially through its relationship to the vector equilibrium, the 24-hour clock and the primes being predictable, something thought impossible up until now.

On March 20, 2019, Cornell University published a peer-reviewed article on their website written by Robert E. Grant and Talal Ghannam entitled, "Accurate and Infinite Prime Prediction from Novel - Quasi - Prime Analytical Methodology," which confirms the ability to predict any prime number.[40]

Grant explains all this in detail in a video he published from a Resonance Talks presentation, "One is the Only Constant," on YouTube.[41]

The True Planck Length

Grant, through his work in vortex based mathematics, describes -1 equaling dark and gravity, -3 as the electron, +6 as the proton, 9 as the neutron, with a neutral charge of 0, the pulse of electricity as -7 and +2, the poles of magnetism as -5 and +4, and +8 as photon and light. Gravity is controlled by 147 and electromagnetism by 852. These are explained in further detail on his website. He continues in the video by saying,

"One of the big problems in physics is we've not been able to know what the real value of the Planck length is. It's irrational. Here's how you get it. The square root of 10, plus 3, divided by 10, plus 1, times 10 to the -34th power gives us the perfect Planck length for the first time:

*"Lp (Planck length) = ((10.5)+3)/10)+1)*10-34 (meters)*

*"= 1.616227766*10-34 (meters)*

"A Sims physicist, Oliver, tested this to come up with the right value for gravity, and it worked because we don't know the exact value of a Plank length but now we do." [42]

A Planck unit in particle physics is a system of units of measurements defined in four universal physical constants of G, C, H, and KB. It describes the measurement of space theorized to be the smallest length possible which is about $1.6*10-35$ meters in the scientific community. Grant states that it was derived from understanding the constants on the Prime Wheel. Vortex based mathematics and harmonic math is solving life's biggest mathematical problems in such an elegant way.

The Hebrew Name of God and the Divine Angle

According to Grant, the musical notes, their hertz value and their relationships in geometric form, are all within the cuboctahedron structure. At the very center of it are 8 points going out at 137.5°, which is the divine angle in Phi.

The following is a diagram Grant derived from using the constants of the Prime Wheel in Phi. He shows us that Phi is the basis for the development of all mathematical constants.[43] The Rodin Phi Diagram I discovered, shown in Chapter 2, is another example of Phi at the point of emanation or point of unity seen in the Rodin symbol. This is the point where all energy comes from and is source energy.

Image 53: Fundamental Constants Mirror
Transform Infinitely in Mirror Symmetry
by Robert E Grant (following page)

FUNDAMENTAL CONSTANTS MIRROR TRANSFORM INFINITELY IN MIRROR SYMMETRY

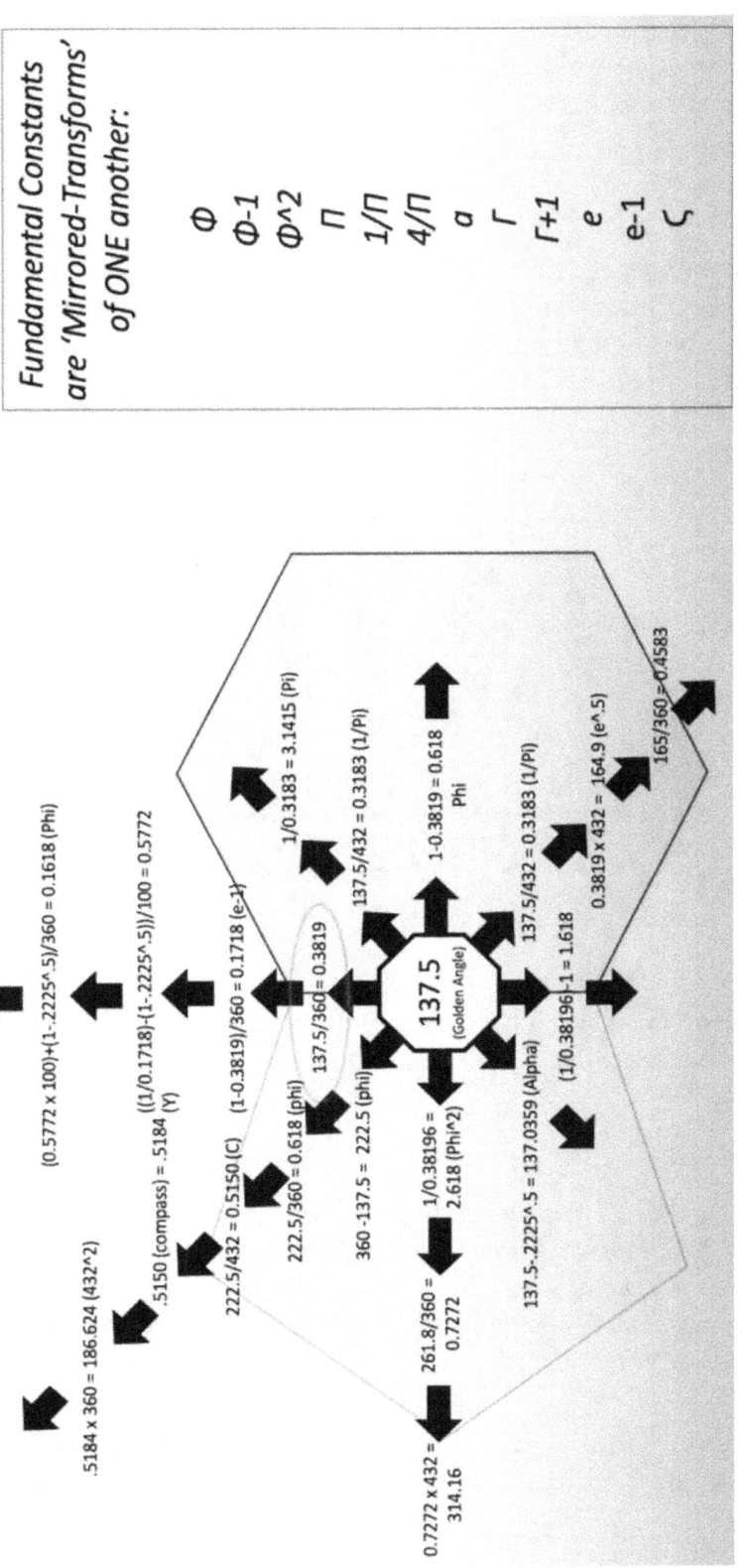

"It's one of the greatest damn mysteries of physics: a magic number that comes to us with no understanding by men. You might say the 'hand of God' wrote that number and we don't know how he pushed his pencil."

~Richard Feynman, theoretical physicist and particle physicist, *"The Mysteries of 137"* [44]

This divine angle of 137.5° is also encrypted in the Hebrew text for the name of God, "Yahweh." Grant states, "It relates 137 to the true only constant which is Yahweh, because Yahweh is actually Pi times seven over Pi to the seventh power." Pi x 7/Pi^7 = 22/3020 which reduces to 1/137.

Image 54: The Hebrew Name of God

The name of God, in Hebrew, seen within the letters as a numerical equation is the divine angle's numerical value as the denominator of 1.

A Musical Matrix

Grant states that, *"Every one of the mirror reflections are fundamental constants. This is really, really important to note and understand that this (vector equilibrium) then infinitely propagates because geometry is sound suspended, matter is light suspended, and it is musical. This is a musical matrix. All the 9-based mathematics that gives us geometry gives us relationships to all these math constants, and the periodic table of elements."*

Furthermore, Grant gives us his Precise Temperament tuning within the cuboctahedron, derived from the vector equilibrium of the Prime Wheel constants.[45]

Image 55: Precise Temperament Musical Geometry
by Robert E Grant (following page)

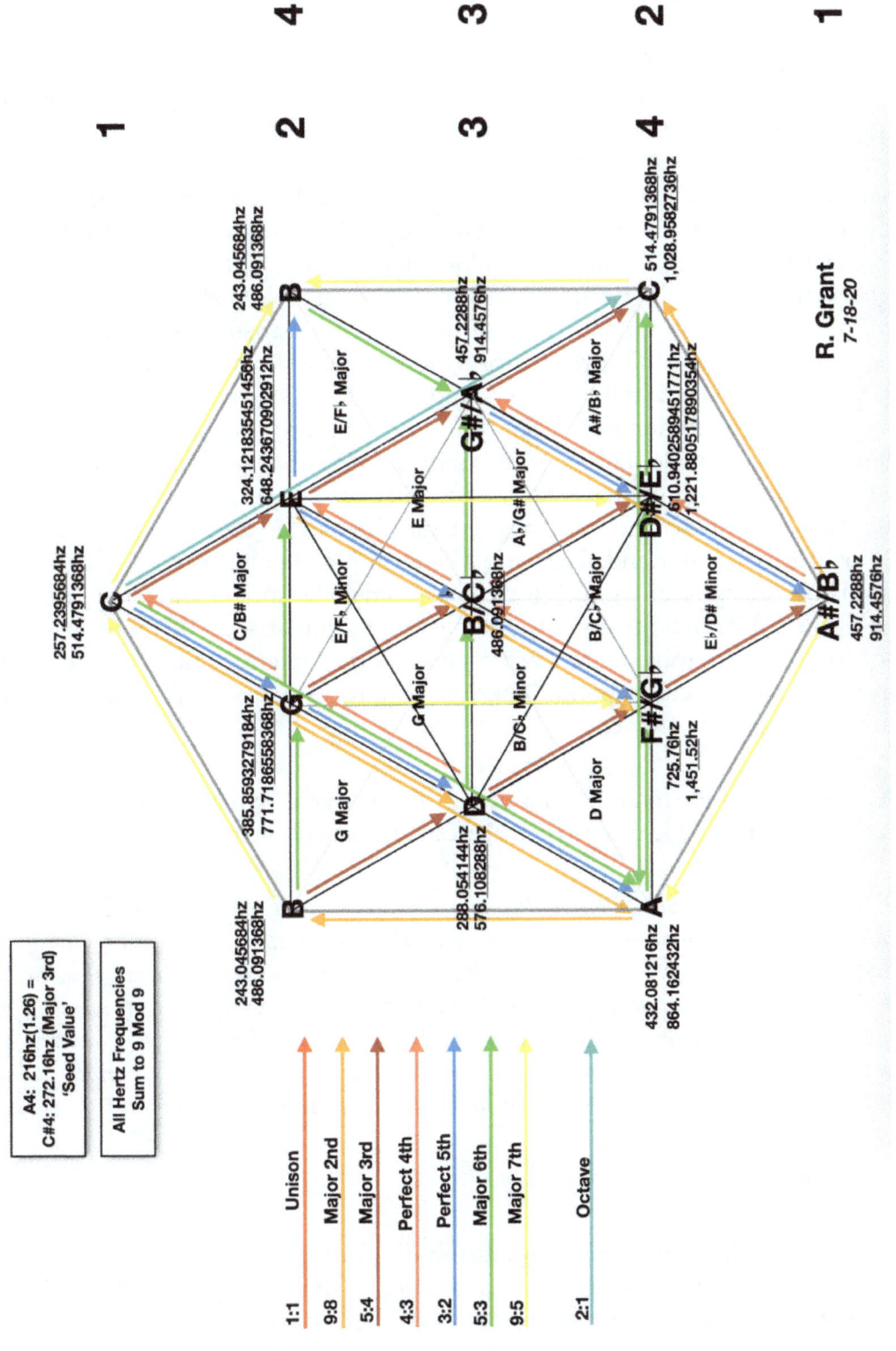

Malcolm Bendall and the Thunderstorm Generator

Future technology will be conscious technology and many technologies will be "tuned" like musical instruments. Would you like to see one of these tuned instruments? Although this next scientist can't quite make his motor sing, he states that for optimum performance, it must be tuned like a musical instrument with precise geometry and resonance. Malcom Bendall is a scientist from New Zealand that has discovered a way to harness the energy of plasma through studying ball lightning emitted from tornadoes and thunderstorms. His Thunderstorm Generator, part of his Molten Sea Ark Atomic Reconstruction Technology (MSAART), creates plasmoids which when run through an internal combustion engine running on gasoline, transmutes the carbon dioxide, carbon monoxide and other hydrocarbons to clean and breathable oxygen.[46] He states that it wasn't until he started to understand the nature of time and matter in relation to harmonic math and numbers, seen throughout nature and our solar system, that he realized the proper angle, ratio, and numerical harmonics to make it work. These are the same harmonic numbers mentioned in this book, and much of the information on the astrological ratio and numerical harmonics can be credited to the work of the well-known geometrist, Randall Carlson. For example, Bendall uses a 4/3 and a 3/2 ratio within his nested spheres capturing the plasmoids. He understands its ratio harmonic as well as the 4 3 2 relationship to the radius of the sun at 432,000 miles. His Plasmoid Unification Model graphs out all harmonic numerical relationships and their attributes within a beautiful circular graph. There is way too much detail to add it here so I recommend finding it online.

Bendall also understands the nature of the toroidal dynamics within the creation of matter and energy. He describes plasmoids in the following quote from his website:

"Plasmoids are doughnut or toroidal shaped clusters of net Protons or net Electrons that once captured and placed into a toroidal orbit are capable of absorbing, storing, and releasing enormous amounts of energy present within their self-generated and structured electromagnetic containment field. Plasmoids, in effect, function as an atomic battery that can be self-charging due to its ability to convert matter to available clean energy. Plasmoids, by their unique geometry, cause a consequential electromagnetic containment field to generate a Zero-point naturally and casually, without much effort, and have the ability to convert the nuclear mass of protium (hydrogen atoms) into energy."
~Malcolm Bendall [47]

Also attributed, Bendall credits Marko Rodin for his understanding of vortex based mathematics, and how it defines the harmonics of numbers seen in their digital root sums. In fact, sometimes they collaborate with each other on their ideas.

There is so much more I could get into with the work of Bendall, but it would take volumes to explain. I encourage you to see the evidence of his Thunderstorm Generator and Plasmoid Unification Model by visiting the Strike Foundation's website, as well as watching the research podcast by Jordan Collin called, Alchemical Science, on YouTube.[48] He explains the work of Bendall, and Marko Rodin in beautiful and descriptive videos that do more justice than I can do on paper.

Below are a few images of retrofits on various engines, including an industrial exhaust retrofit. His technology is causing many scientists to scratch their heads because of how it works. But the tests by all independent researchers get the same results; it turns carbon gases into oxygen! This technology alone is about to clean the air on this planet, as well as increase fuel efficiency by 15% to 30%. [49]

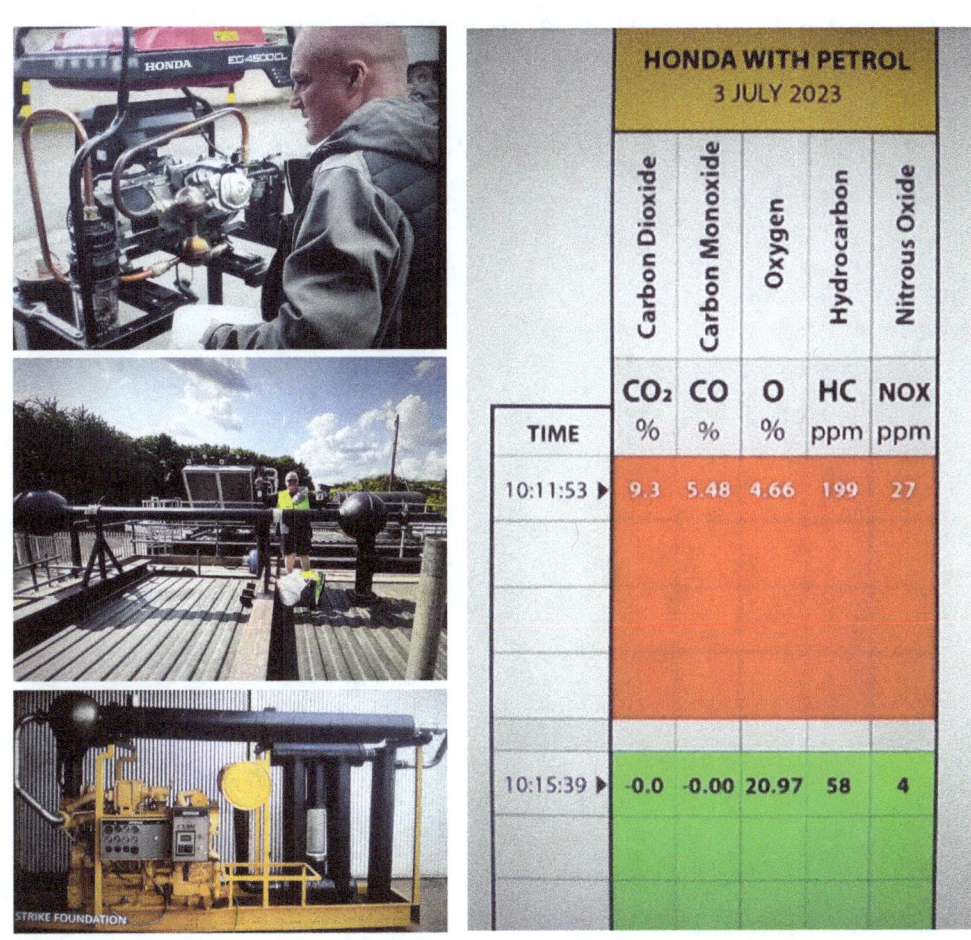

Image 56 – 59: Thunderstorm Generators and Exhaust Output Reading

Bob Greenyer and Ball Lightning

To further validate the effects of the Bendall Thunderstorm Generator, Czech researcher and scientist Bob Greenyer, from the Martin Fleischmann Memorial Project, gives an interesting analysis of what is occurring inside the reactors that are generating plasmoid ball lightning. Being one of the world's leading experts on ball lightening, he has compared various test samples from many reactors and they are all showing the same effects and geometry, including those from the Thunderstorm Generator. The geometry is within harmonic ratios seen in nature and is what he calls evidence of a "toroidal moment."

The following images are some of the examples showing this geometry in the embedments of these reactors. They were analyzed through a scanning electron microscope (SEM) by Greenyer and his team, much of the time during a live online broadcast.

Image 60 – 61: Greenyer Sacred Geometry (following pages)

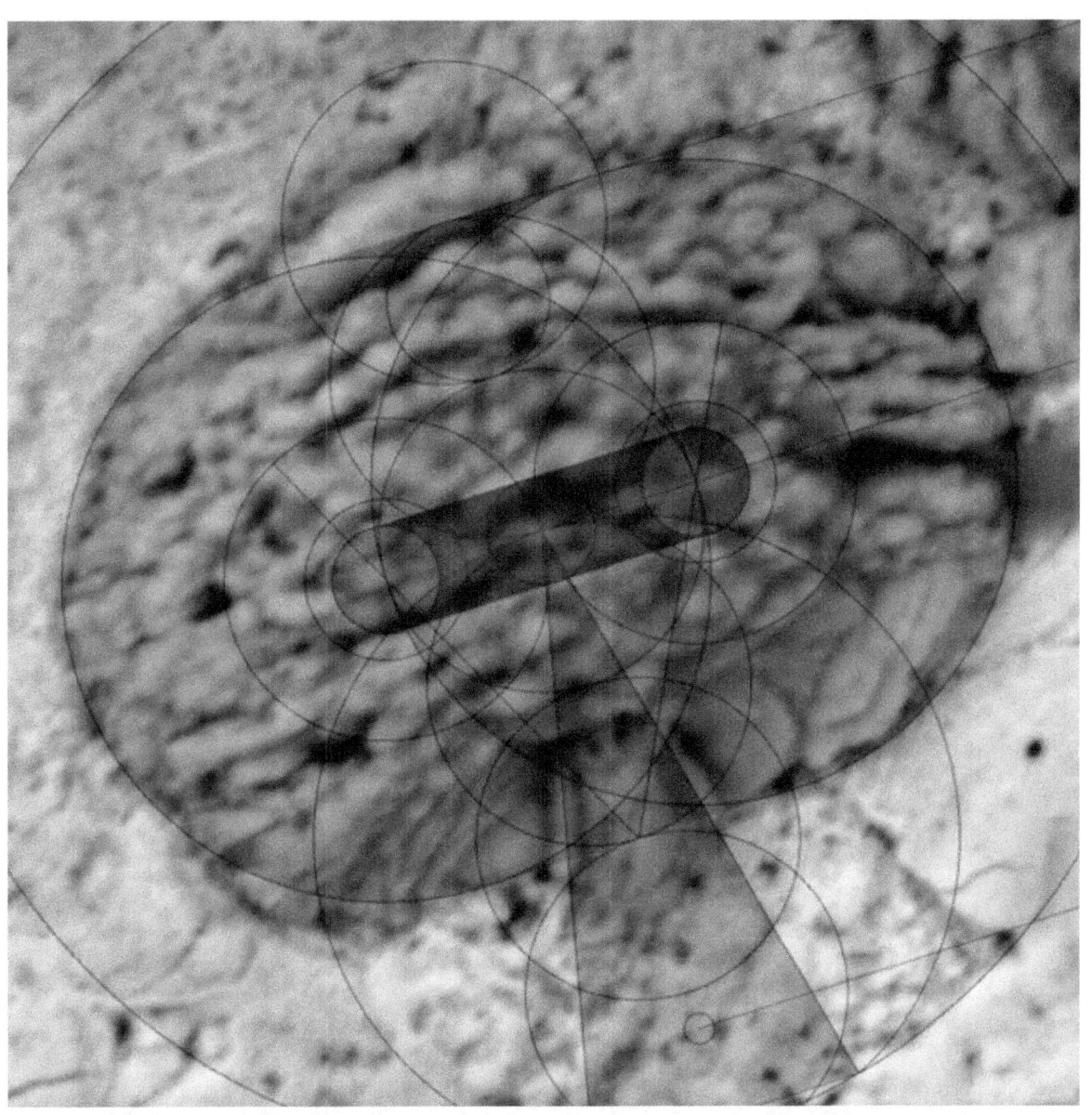

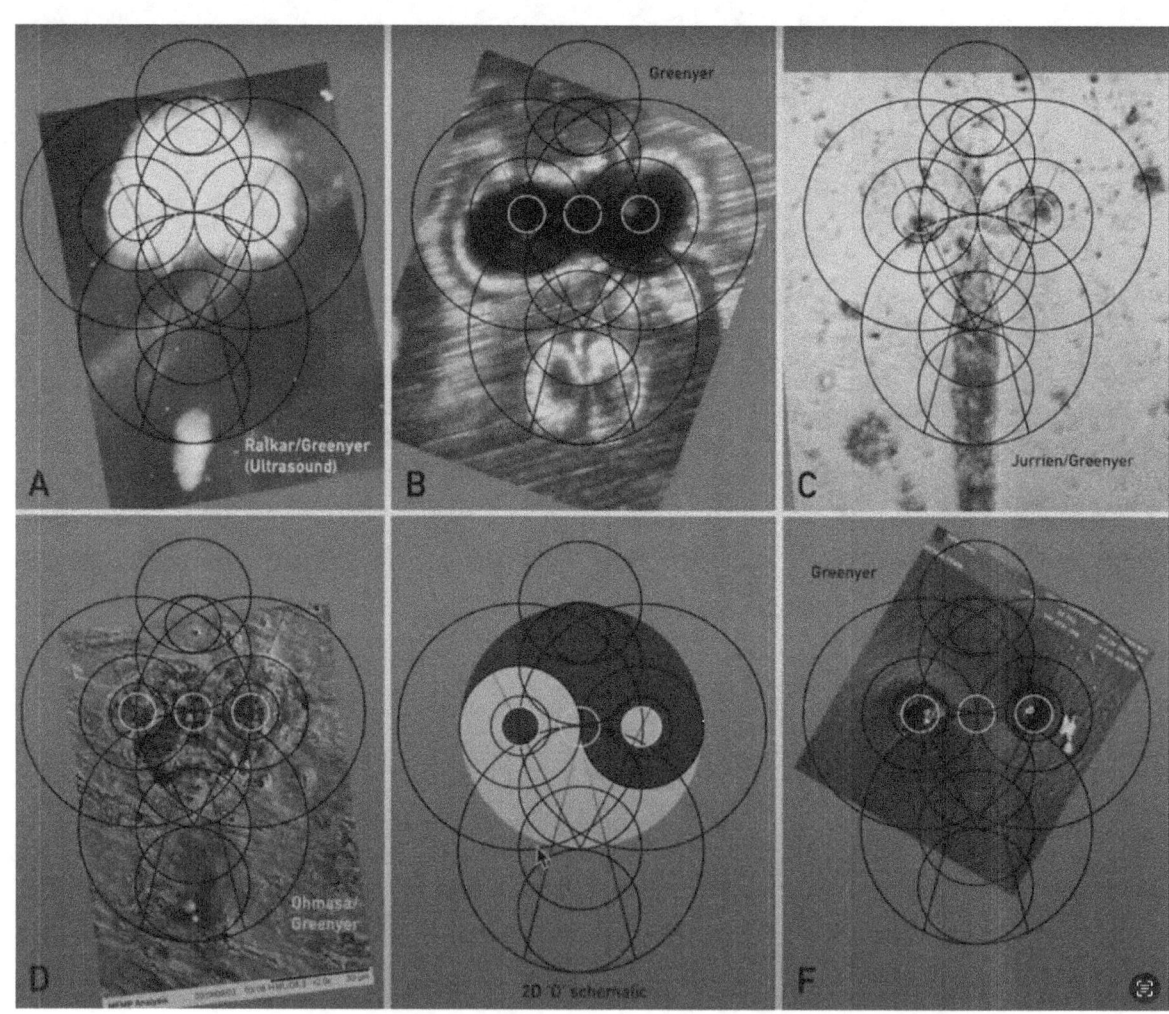

When plasmoids are created and interact with their surrounding materials, an imprint is also created of what is left over like a shadow of its actions. After creating his initial geometric diagrams, Greenyer was shown the sacred geometry by others they had found inherent in them. Greenyer just outlined what he saw within the samples and it just so happened to be a Phi ratio diagram from top to bottom, and also shows the male and female aspects of creation coming together in a process of birthing new elements not present in the material prior to the experiments.

Through his research, he discovers the evidence of this balancing act of destruction and creation. Not only do we see and learn its geometry, we see evidence of the transmutation of elements. Traces of synthesized elements, such as silicone, aluminum, calcium, silver, magnesium, and even titanium have been found in the initial tests. These are not elements that we should typically see present within the makeup of the iron-rich walls the plasmoids were embedded in. He states the samples have, "a spread spectrum of elements typical to stellar synthesis." This is this discovery of the transmutation of elements![23] Again, this is alchemy, pure and simple, and it is not pseudoscience or occult magic, it is real transmutation like what we see in the sun. Continued analysis is further nailing this to the wall of truth in what we are seeing. It is remarkable.

Another name for this process and the geometry we see is what Greenyer calls the yin yang process. We see the yin yang symbol in the valleys and mounds, as well as the opposing small circles as the creation and destruction points where elements are torn apart to structure new elements. They perfectly lineup on the dot, literally. Where the small black and white dots would be in any yin yang image, they are seen clearly in the following sample.

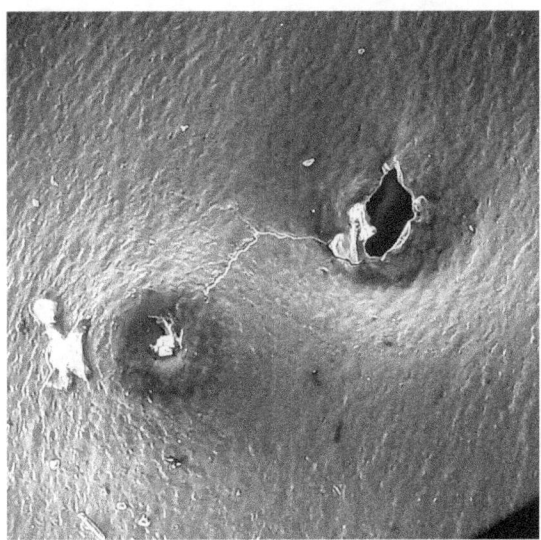

Image 62: Greenyer SEM Yin Yang Structure

The center points within the two small circles that mirror each other at the top of the Greenyer, yin yang sacred geometry is a side view of what a toroid shape would look like if cut in half from the top down the middle, and looking from its side. This confirms the toroidal vortex action that must take place at any point of the creative process of tapping into source energy at its point of origin. Again, what Greenyer calls a "toroidal moment", and what Rodin would call the point of unity origin.

Later I noticed that the lower triangular part of his diagram looked to be the same as another I came across in my research. This resulted in an interesting discovery of my own.

Charlie Ziese and Universal Phi Scaling

Bob Greenyer's sacred geometry analysis of plasmoid embeddings, like the one seen in the Thunderstorm Generator and other reactors, precisely matches the universal Phi scaling, discovered by Charlie Ziese.

In his book, *"76.345 – Exploring the Hidden Secrets of the Golden Ratio,"* evidence of the lost knowledge of Phi scaling in energy implosion theory is revealed in ancient structures found all throughout the world and all throughout history. It is the angle of the 76.345° spires we see towering the rooftops of many historic churches and cathedrals, as well as many other structures seen extensively in his book. It is also very close to the Alexander Golod Russian pyramids.

This angle is also seen in nature, for instance, through the mirror refracting in the way light travels as it enters our eyes and flips the image in what is referred to as the anatomy of spherical refraction. It is also the angle seen in a spiraling whirlpool vortex or tornado, and is also used in the angle of acoustic sound amplification. The following image gives us its basic layout.

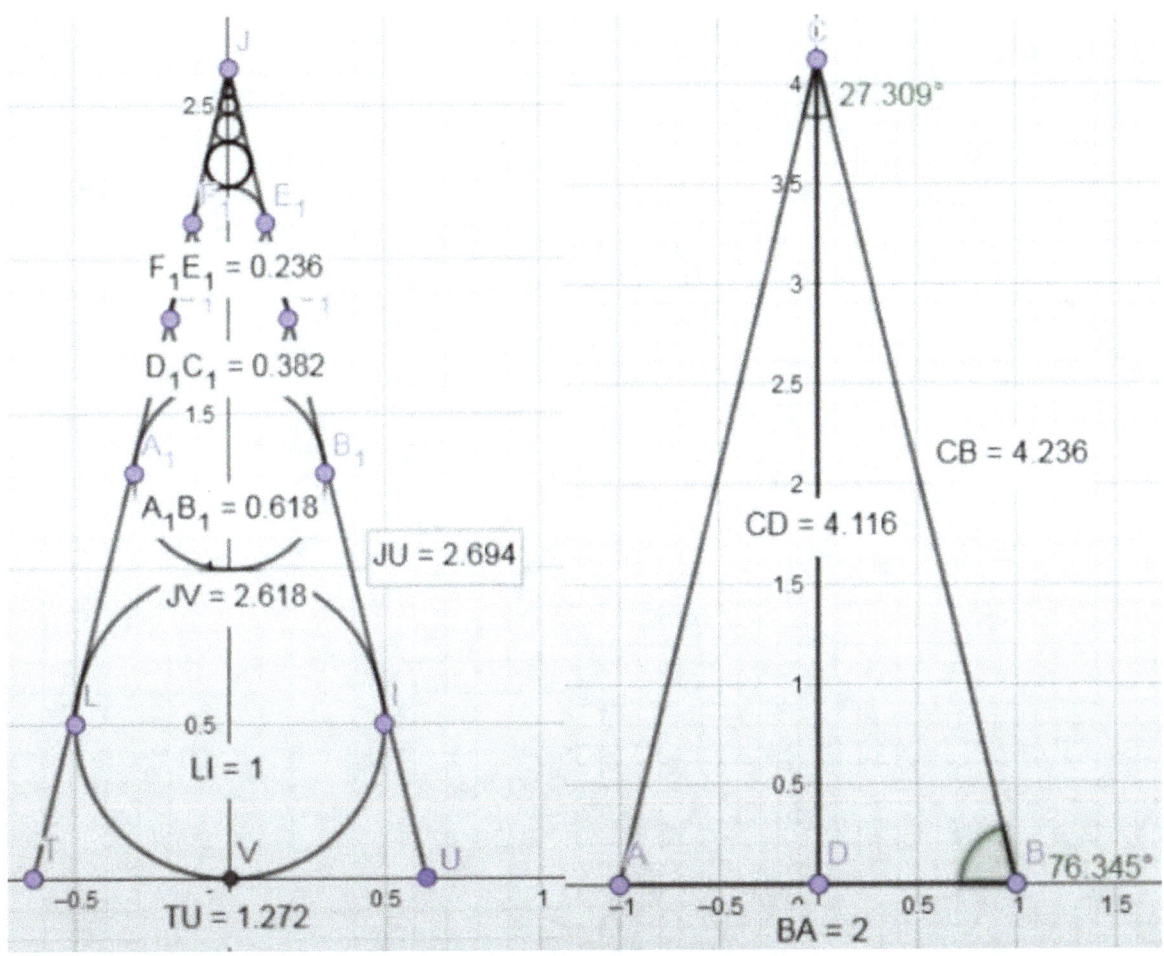

Image 63: Ziese Universal Phi Scaling Model

Scaling a set of spheres according to the ratio of the golden mean gives us a scaling degree of implosion or expansion of 76.345°. It is a lost golden angle based on spherical scaling. If the largest bottom circle has a diameter of 1, then the sequential sphere above it would get smaller at a ratio of 1 to 0.618. So, 1 divided by 1.618 equals 0.618, then 0.618 divided by 1.618 equals 0.382, then 0.382 divided by 1.618 equals 0.236 and so on.

The angular reduction in ratio creates a slope angle and it matches Greenyer's sacred geometry almost exactly. I drew up the following diagram showing the relationship between the two.

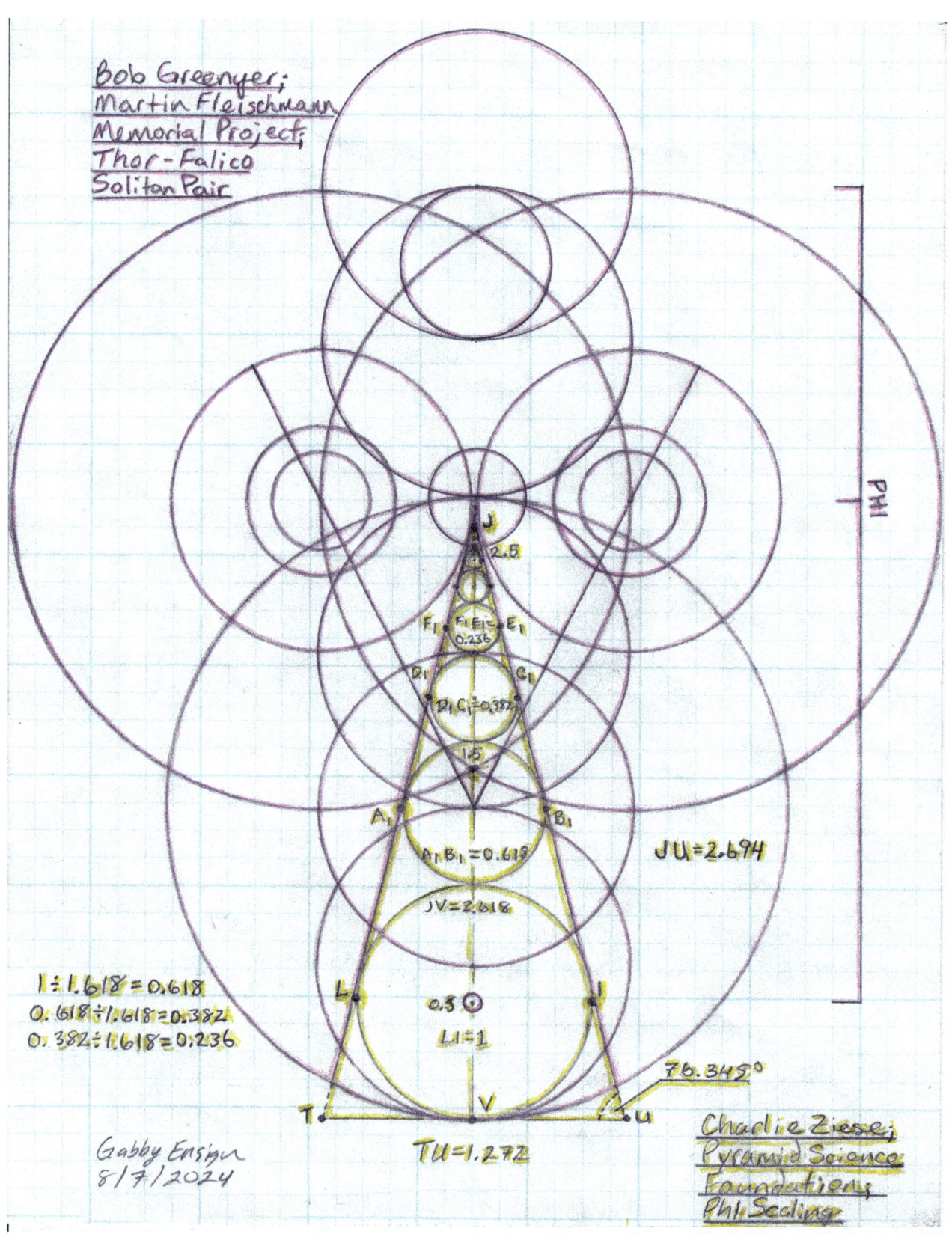

How interesting. The math and the geometry tell us that this precise triangular angle of 76.345° is likely not a coincidence considering the already well established relationship Phi has within the creation process. This concludes that universal Phi scaling IS a function within the geometry of plasmoids during its stages of transmutation, and must also be evident in the phenomenon of ball lightening and vortex/energy implosion technologies. This scaling ratio may lead to further discoveries in the replication of nature's most useful properties.

I shared my diagram with Bob Greenyer around mid-August 2024, and after a detailed and grateful response explaining and giving credit to those who discovered the geometry of Phi seen elsewhere in his diagram, he thanked me for all of the insights I had. He stated that my addition may be a more accurate depiction of the actual geometry then that of his observed approximations and he would give me credit for that.

The credit really goes to Charlie Ziese and Dan Winter for rediscovering and teaching universal Phi scaling. I and others just discovered the relationship of that to the work of Greenyer.

Interestingly, the following evening after returning home from work, I brought up YouTube on my TV to see if there were any new videos from my subscriptions. The first video I saw was Greenyer's most recent live podcast title, *"O-Day – Relentless."* The thought in the back of my mind at that moment was, "I wonder if this video has anything to do with my contact with him the day before." It didn't take but a few seconds to realize it was, when I saw numbers on the title image like 0.618, 1.618, and 76.345; and phrases like "working with nature."

I watched as he discussed the concepts he had recently learned about which included a new irrigation technology and plans to introduce and push forward their ideas for minimum viable products (MVPs). These would be affordable and effective technologies utilizing toroid and vortex physics, primarily through plasma and harnessing toroidal solitons. About an hour and a half in, he started to show and discuss my diagram and gave credit to those who also shared their revelations regarding the Phi aspects and the sacred geometry it displays.

He credits Robert Hutchings from Hawaii for a similar Phi depiction, but not exact, and containing other Phi relationships. He credits a project follower called Tony Jabony, who found more Phi relationships within his geometry. Not mentioned in the video, but in my email exchange with Bob Greenyer, he gives a respectful reference to Dan Winter for "his ideas around Phi ratio vortex collapse." It is an honor and a joy to be able to contribute to any part of understanding the process of creation. It just so happens to be Phi, again!

About a week later, Bob Greenyer hosted another live broadcast where he discovers within his CAD software that there are two different angles to measure the Phi scaling of spheres; by the tangent of the spheres and by the diameter of the spheres. Each one will result in a different angle. The tangent is 76.345°, and the diameter is 76.72°. Robert Hutchings shows the latter. He argues Hutchings is most likely correct in the construction of the cone geometry seen in these structures, whereas I believe the 76.345° angle is seen in square pyramid geometry structures. This is because a four sided pyramid with equal sides and a slant angle of 76.345° will always have a base angle at its face of 76.718°. Rounded up gives us 76.72°. There is a relationship. I discovered this while researching the angles being discussed. Using either angle would probably give the same results because they are so close to each other and Phi will "bend" to anything within close range of 76°.

Vortex Energy Implosion Technologies

We can see that there is a vortex implosion of energy within this model of physics. Not explosion, like we have been trying to master for so long now. Vortex energy technologies are designed and are a reflection of nature's way of doing things. And that is the best and most effective way nature could think to do it. It's perfect. Anything we do or make in this regard from this point out will be basically useless unless it utilizes the laws and design of nature, and it's done so with harmless intent. Conscious technology. The following example is another demonstration of what conscious technology can do.

The Rodin Coil Over Unity Transducer

Another example of a tuned device is the Rodin Coil. Marko Rodin and a colleague of his, Randy Powell, created a device that adheres to the vortex based mathematics concepts and instructions called the Rodin Coil Over Unity Transducer.[50]

The first Rodin Coils were made around the year 2009 I think. It has been a challenge to determine the right windings, frequency, and wave alignment to get the best results. Copper has been the primary material used due to its conductivity.

There are many videos online showing people getting some results, others getting amazing results and then those with no results. It is not as cut and dry as you would think. A few claim to have been able to produce over unity, which means they get more output voltage than what they put in, but they make it difficult to see or understand their method or set up. Others show you "tricks" like playing music through it like a speaker, levitating an iron ball, or spinning an iron ball in a glass dish that has been charged up.

After a while of seeing such small gains in voltage, if any, I became a little discouraged about the Rodin Coil being correctly designed and built. I believe the technology is there, but more research is needed to advance this technology.

That being said, one individual is showing phenomenal results. In a live demonstration in front of an entire conference of people, presented by Marko Rodin, Russ Grease, and someone who went by the alias "007" (because he was afraid he could be targeted like so many other inventors of breakthrough free energy devices), positive results are being shown. The alias, "007" never showed his face in his live video presentation. The conference was held by Global BEM (Global Breakthrough Energy Movement), a nonprofit, in 2016. The entire demonstration can be found on YouTube; *"The Rodin Coil Over Unity Transducer,"* and is about two hours, so if you would like to go straight to the input versus output demonstration it starts at 1:21:30.[51]

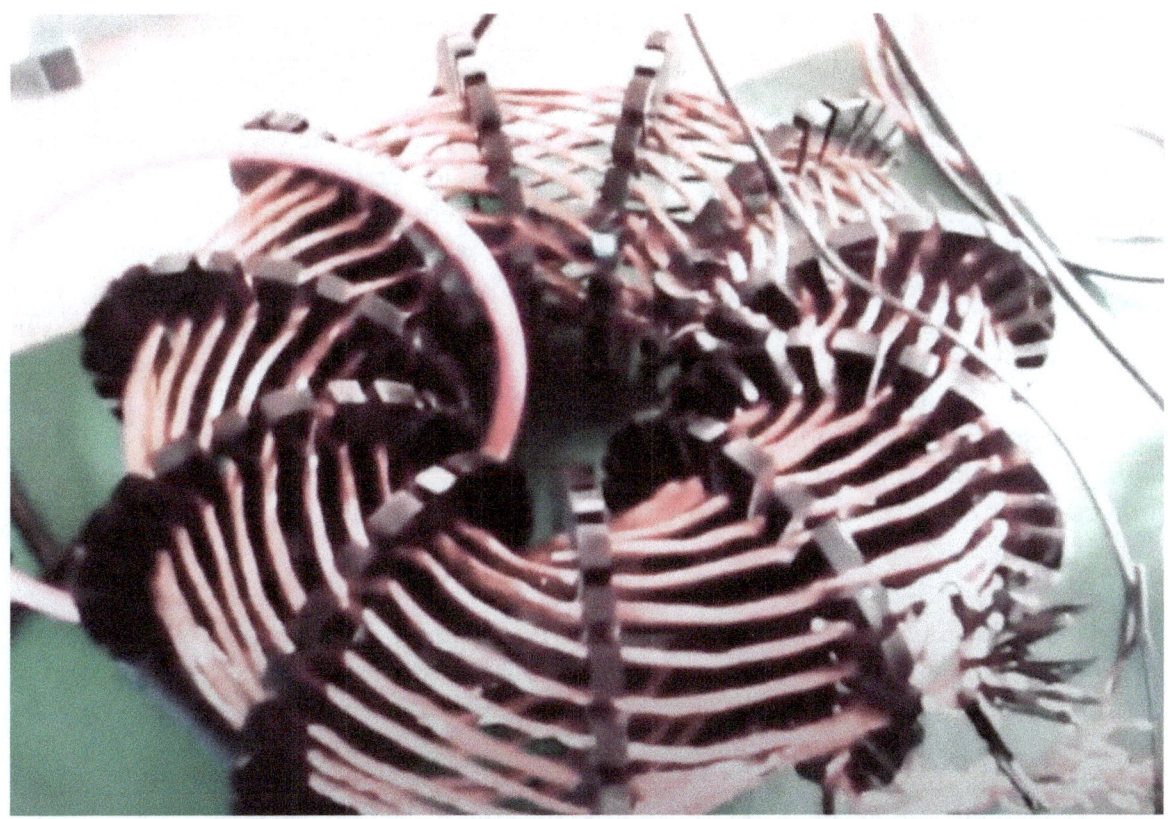

Image 64: Rodin Coil

It was 007 who developed this latest design and it had to be "tuned" properly to work, just like a musical instrument. The sine wave frequencies between voltage and current needed to be adjusted through a phase shift in an oscilloscope to the proper lag distance and interestingly looks like the 1 2 4 8 7 5 circuit of the Rodin symbol. It also had to be tuned to 1.226 khz, which I don't fully comprehend, but future experiments may tell us. If the input signal went out of phase shift or the hertz frequency strayed from 1.226 khz, the light bulb array it was connected to would dim significantly. His settings had it putting out 74.3 VAC and when the frequency was dropped to 1.026 khz the bulbs nearly went out. The device was going out of resonance because resonance is key in all of creation. Why would it not be in a device designed to harness this same energy?

The input into the coil came from a bass amplifier and was sending 37.3 VAC at 1.15 Amps and was putting out 110 VAC at 1.112 Amps. That is a 294.3% increase in power!

The coil itself was about 10.5 inches x 3.5 inches with about a 3.5 inch hole in the center. Weirdly it operates on reduced induction and even drops in temperature within the center of the toroid, instead of increasing in temperature, which is calculated as losses through its radiation and dissipation of energy as heat. The reason it drops in temperature is that it is creating a zone of low pressure within the very center compared to its surrounding vortex and as a result produces lower temperatures closest to its center. That is exactly what you see happening in the center of a tornado. Well of course, that's what toroids do, unless it is a black hole, but we still have yet to fully understand those. The oddities and curiosities continue. The ohms resistance decreases, even though the voltage increased and the current in amps stayed about the same. The opposite is true, using any other method of electrical amplification. And lastly when 007 ran the oscilloscope from the input displaying the dual phase shifted sine wave signal (resembling the Rodin Symbol) to the output running between the coil and the lightbulb display, three sine wave patterns emerged, something 007 had never seen before in his experience with electricity. The energy of creation is of 396, so 3 waves sounds like an accurate statement to me.

Russ asked 007 if he thought changing the material from copper to something else would increase its efficiency and he said using carbon graphene would, having an atomic weight of 6. Using Ohm's law, if you increase the temperature on a conductor, the resistance will go up, which is why electronics need to be cooled by fans because of the resistance within the circuitry and wiring. The incredible thing about the graphene was that when 007 heated it, the ohms dropped from 1.5 ohms to 0.7 ohms. He stated that if you used it instead of copper, you would increase the efficiency dramatically. He demonstrated this by hooking up a small 1" x 4" thin piece of graphene to a multimeter and heating it up with his lighter. You could see the display showing 1.5 ohms then going down to 1.1, then to 1.0, 0.9, 0.8 and finally 0.7 ohms before he burned himself and bumped the graphene off the multimeter. With graphene, 007 theorized that once you started it with an input, you could disconnect the input and it would run on its own. Marko asked if it was possible that it could be a closed-circuit device at some point, meaning it would not require an input to start it, because Marko's math showed that it was possible. 007 replied at some point yes, but it would require more advanced nanotechnology.

The reason this technology works and demonstrates properties and anomalies like it does is because it is harmonically in sync with the same energy that is the source of the non-decaying spin of the electron. Nikola Tesla was correct about tapping into the wheel work of nature, and understanding that the secrets of 3, 6 and 9 would give us the keys to the universe. We just have to fine-tune this technology to see what it is actually capable of. This will happen and is already happening with independently funded research.

Sheela Rahman and the XYZ Oscilloscope

Imagine a stone being dropped into a pond. See the ripples as they move across the surface of the water, like little waves about to come ashore. Those ripples are akin to any frequency you might see elsewhere in nature. A frequency is a wave pattern, in this case, a ripple on the pond. Of course, sound does the same thing and like cymatics, carries its form in 3-dimensions and can be plotted on an XYZ graph.

Sheela Rahman is a sound engineer, musician and frequency researcher. She has discovered an amazing way to test the coherency of any frequency by examining the nature of the wave patterns generated using a 3-D oscilloscope.[52] The three planes of the XYZ axis are variables using frequency (or pitch), amplitude (or voltage) and time (or phase angle). This method gives us a way to see these frequencies in their three-dimensional form and they appear as geometry. Cymatics literally only touched the surface. Sheela has turned it up a whole notch to 11, and gives us an incredible look at what harmonic frequencies look like when compared to those that are not. Harmonic frequencies show stable and coherent geometries, while others that are not show erratic and incoherent behaviors that never create any form of geometry.

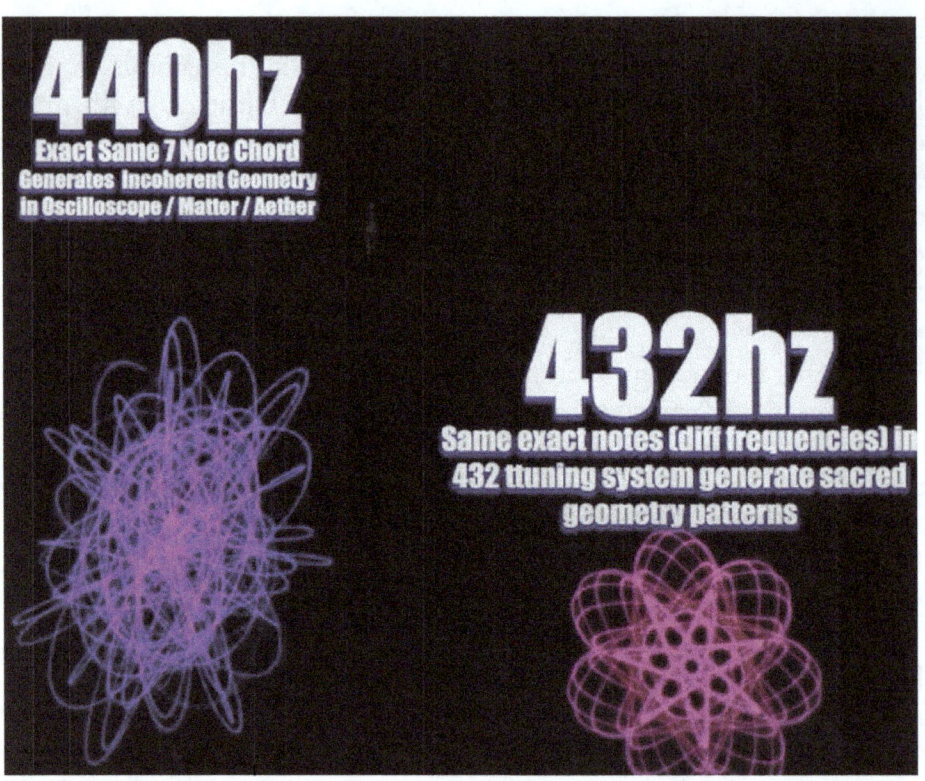

Image 65: Sheela Rahman 440hz and 432hz Comparison

Sheela's image is a great example of this because of the implications of how we can use this tool. Tuning our instruments to A432, like some cultures have done in the past, will result in music that mirrors the same coherent geometries seen in nature. This can only be a good thing.

She put this to good use. Sheela took the numerical lines of the Malcolm Bendall Plasmoid Unification Model, and inserted them into her oscilloscope. When she combined all of the numbers of any given line as hertz frequency into the oscilloscope, they would all show stable and coherent geometries. Even combining all frequencies of all numerical lines would still maintain a resonant field as a grand stable and coherent structure. Any deviation using random frequencies would shatter the stable geometry. Here are a few examples of some of the harmonic tones she was experimenting with.

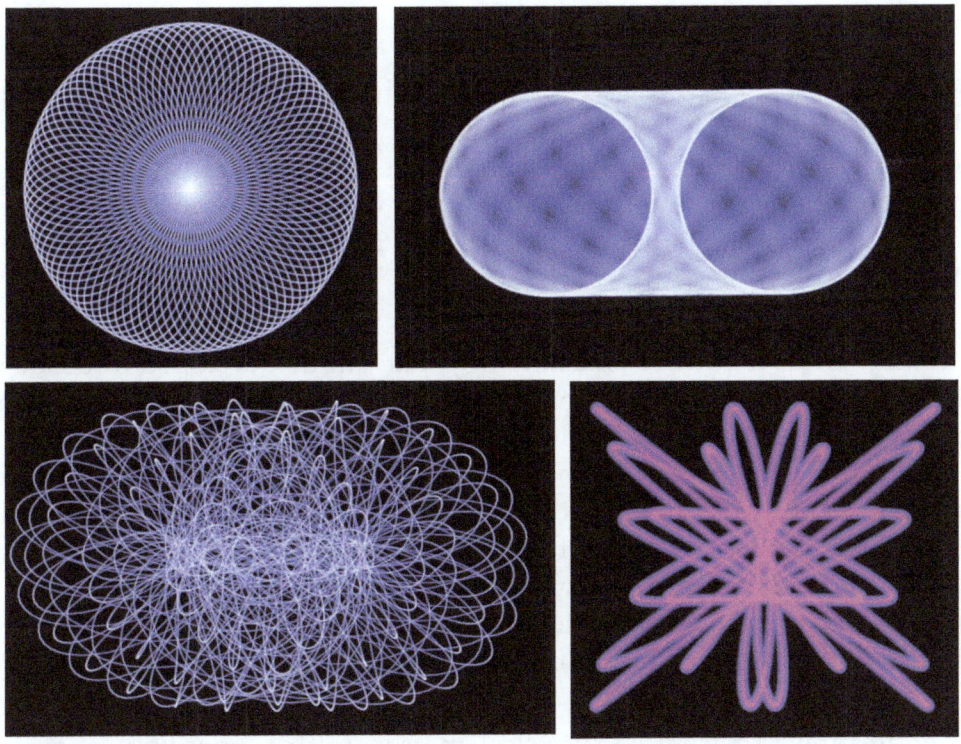

Image 66 – 69: 174hz x 396hz, 741hz x 963hz, 396hz x 963hz x 639hz and Line 15 of the Malcolm Bendall Plasmoid Unification Model

So what are the frequencies derived from the Plasmoid Unification Model? They are the same as what we see all throughout this book. The same numbers as those seen in the chart of most harmonic numbers. The same as the angles of the major polygons, the distances and sizes of planets, and the notes in the A432 tuning system. They are even displaying the 4/3 and 3/2 ratio relationships. And yes, they all add up to a root sum of 9.

Sheela Rahman gives us a great tool to see just how significant Malcom Bendall's model is. It appears to be completely in balance. She even uses a Rodin coil as a broadcast antenna to send ripples of harmonic fields into the room, because the Rodin coil amplifies its most natural path and creates an electromagnetic field which our bodies can detect. There is much that needs to be healed.

The Keys to a Better World

So now that you can see these potentials within your physical world, understand the holographic message it tells us. Quietly tune in to the source of all, the divine, through the right frequencies of love and forgiveness, and you shall move into sync with a source of all power, and then at that point, we are fully connected to the divine, the true source of power. Those are the real keys to the universe if we choose to take that path.

Otherwise, know that there is technology that can be developed to create a better world for ourselves. It is not just a possibility, it is a reality, just waiting to be set into motion, whether it is from the Rodin Coil or any other number of possible devices designed to harness energy in clean and efficient ways. The truth is, we are much closer today than we were 100 years ago, so progress is good. But responsibility is a requirement, and love a necessity if we are to advance as a civilization for future generations to come. I guess that is why we have music, just something else to remind us to enjoy ourselves, be kind to ourselves and to each other, and take time to smell the flowers, even if they are ugly, like a sunflower is to some.

The entire universe truly is a musical matrix, full of conscious geometry, and numbers, sound, shape, energy, force, movement, structure, resonance, harmonics, light and love. I can see this by looking at the beauty of numbers behind the beauty of nature. I see all this…

…behind every sunflower.

Gabrielle Ensign

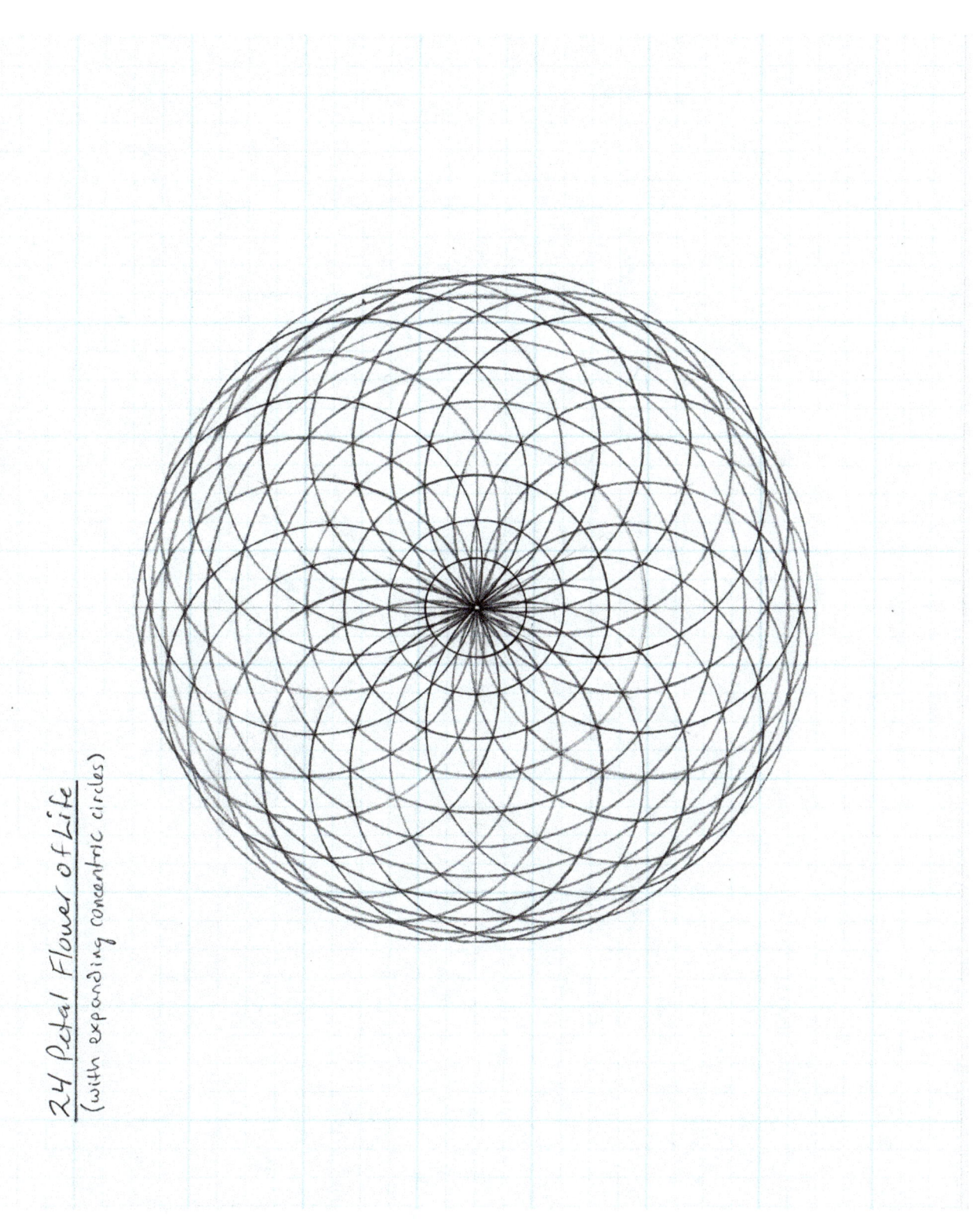

24 Petal Flower of Life
(with expanding concentric circles)

"Nothing real can be threatened.

Nothing unreal exists.

Herein lies the peace of God."

~A Course in Miracles
In: 2:2–4

About the Author

Gabrielle Ensign is an artist and musician who was born and raised in Flagstaff, Arizona. After having an out-of-body experience in 1994 at the age of 14, and having a massive UFO encounter with a group of friends at the age of 15, she was driven to seek answers to life's biggest mysteries.

At the age of 17 Gabrielle discovered The Delfin System by Leslie Fieger. This introduced her to more philosophy, physics, metaphysics and consciousness and it was the first time she learned of Phi and the golden ratio.

At 19 she was immersed in books relating to the subconscious mind including, *"The Master Key,"* by Charles F. Haanel, and, *"The Power of the Subconscious Mind,"* by Dr. Joseph Murphy.

By the age of 24, Gabrielle was well versed in many esoteric, spiritual, scientific and historic topics by authors such as Edgar Cayce, Nikola Tesla, Wilhelm Reich, Walter Russell, and, "The Law of One," by Elkins, Rueckert and McCarty. She has since continued her search for truth in whatever form it takes, which has included many little miracles along the way.

Gabrielle is currently 44 years old and has over 30 years of wisdom and experience pertaining to many of these topics. She spends much of her free time playing music in Sedona, Arizona. You can hear some of her music at SoundCloud.com/BridgeToArcadia.

Image Credits

All gridline diagrams courtesy of the author. Specific credits applied to image.

Image 1:	Courtesy of the author, credit Marko Rodin
Image 2 – 10:	Licensed from Envato Elements
Image 11:	Credit Marko Rodin
Image 12:	Courtesy of the author
Image 13 – 22:	Credit Gary Meisner
Image 23 – 24:	Courtesy of the author
Image 25 – 27:	Credit Hans Jenny
Image 28:	Credit NASA
Image 29:	Courtesy of the author, credit Richard C. Hoagland
Image 30:	Credit NASA
Image 31:	Credit Google Earth
Image 32:	Credit NASA, JPL, GSFC, Arizona State University
Image 33 – 34:	Credit NASA, ESA, and J. Nichols, University of Leicester
Image 35:	Courtesy of the author, credit Ivan T. Sanderson
Image 36 – 37:	Credit R. K. Harrison
Image 38:	Credit Joscelyn Godwin
Image 39:	Courtesy of the author, credit E. Battenar, E. Florido
Image 40 – 42:	Courtesy of the author
Image 43:	Courtesy of the author, credit Niles Johnson
Image 44:	Courtesy of the author
Image 45:	Credit David LaPoint
Image 46:	Credit IBM

Image 47:	Credit TEAM 0.5 at Lawrence Berkeley National Laboratory and Berkeley Lab's National Center for Electron Microscopy
Image 48 – 49:	Courtesy of the author
Image 50:	Courtesy of the author, credit Nassim Haramein
Image 51:	Courtesy of the author, credit Luke Ragle
Image 52 – 54:	Credit Robert E. Grant
Image 55:	Courtesy of the author, credit Robert E. Grant
Image 56 – 59:	Credit Malcolm Bendall
Image 60 – 62:	Credit Bob Greenyer
Image 63:	Credit Charlie Ziese
Image 64:	Credit Marko Rodin
Image 65 – 68:	Credit Sheela Rahman

Bibliography

1. National Geographic, Van IJken, Jan, *"See A Salamander Grow From a Single Cell in this Incredible Time-lapse | Short Film Showcase,"* https://www.youtube.com/watch?v=SEejivHRIbE

2. https://markorodin.com; Google search: "vortex based mathematics"

3. Britannica, The Editors of Encyclopaedia Britannica, *"Fibonacci Sequence,"* 2024, https://www.britannica.com/science/Fibonacci-number

4. Devlin, Keith J., *"The Golden Section: A History of Philosophical Ideas,"* Chichester: Wiley, 1990

5. Meisner, Gary B., https://www.goldennumber.net/golden-ratio/

6. YouTube, @inphiknitfractal, *"Gator 18x18 Grid - Vortex Mathematics - Sacred Geometry,"* https://www.youtube.com/watch?v=h3S-DmqJ_zY

7. YouTube, @whyphi?, *"Ever Heard of The Root Fibonacci Series? | Advanced Vortex Based Mathematics,"* https://www.youtube.com/watch?v=xYGSs5XC-HE

8. Jenny, Hans, *"Cymatics: A Study of Wave Phenomena & Vibration,"* 2001

9. NASA, ESA, JPL, SSI, Cassini Imaging Team, , 2012, https://apod.nasa.gov/apod/ap230618.html

10. Hoagland, Richard C., *"The Monuments of Mars: A City on the Edge of Forever,"* Frog Books, 2001

11. Newman, Hugh, *"Earth Grids: The Secret Patterns of Gaia's Sacred Sites,"* Wooden Books, 2008

12. Grant, Robert E., https://robertedwardgrant.com/precise-temperament-tuning

13. YouTube @NFGC Tarot, *"Vortex Based Math & the Flower of Life Discovered & Copyrighted by Bill Wandel,"* https://www.youtube.com/watch?v=kGcAyMaSyXU

14. Britannica, The Editors of Encyclopaedia Britannica, *"Metric System,"* 2024, https://www.britannica.com/science/metric-system-measurement

15. Roberts, Sarah, *"Babylonian Astrology: How Mesopotamian Priests Influenced Your Horoscope,"* 2018, https://www.ancient-origins.net/history-ancient-traditions/babylonian-astrology-0010806

16. Britannica, The Editors of Encyclopaedia Britannica, *"Swastika,"* 2024, https://www.britannica.com/topic/swastika

17. Harrison, R.K., Journal of the Evangelical Theological Society (JETS) 36/1 3-8, *"Reinvestigating the Antediluvian Sumerian King List,"* 1993, https://etsjets.org/wp-content/uploads/2010/07/files_JETS-PDFs_36_36-1_JETS_36-1_003-008_Harrison.pdf

18. Ibid.

19. Godwin, Joscelyn, *"Atlantis and the Cycles of Time: Prophecies, Traditions, and Occult Revelations. Inner Traditions,"* 2011, pp. 300–301. ISBN: 9781594778575

20. Britannica, Van Buitenen, J.A.B., Lin, Chao, The Editors of Encyclopaedia Britannica, *"Calendar,"* 2024, https://www.britannica.com/science/calendar

21. Coolman, Robert, *"Keeping Time: Why 60 Minutes?,"* 2022, http://www.livescience.com/44964-why-60-minutes-in-an-hour.html

22. Munck, Carl P., Radio Bookstore Press, *"The Code - 1997,"* 1996, ISBN: 188907117X, 9781889071176

23. Ibid.

24. Battaner, E., Florido, E., *"Magnetic Fields and the Large-Scale Structure,"* 1999, https://arxiv.org/abs/astro-ph/9911423v1, https://arxiv.org/pdf/astro-ph/9911423

25. YouTube, @Absalom Melchizedek, *"Vortex Based Mathematics - Marko Rodin,"* https://www.youtube.com/watch?v=fI93jeaXGvs

26. Campbell, Joseph, Princeton University Press, *"Hero With A Thousand Faces,"* 2004, ISBN: 0-691-11924-4

27. Urbantke, H. K., Journal of Geometry and Physics, 46 (2003) 125–150 Institut fur Theoretische Physik, Universitat Wien, *"The Hopf Fibration—Seven Times in Physics,"* https://www.fuw.edu.pl/~suszek/pdf/Urbantke2003.pdf

28. Cockram, Bernice, *"In Focus Sacred Geometry: Your Personal Guide,"* Wellfleet Press. 2003, ISBN:978-1-57715-225-5.

29. Fuller, R. Buckminster, Applewhite, E.J., Macmillan, *"Synergetics: Explorations in the Geometry of Thinking,"* 1975, 1979, ISBN: 002541870X

30. YouTube, @ David LaPoint, *"The Primer Fields Part 1,"* https://www.youtube.com/watch?v=siMFfNhn6dk

31. Gross, Leo,* Mohn, Fabian, Moll, Nikolaj, Schuler, Bruno, Criado, Alejandro, Guitián, Enrique, Peña, Diego, Gourdon, André, Meyer, Gerhard, *"Bond-Order Discrimination by Atomic Force Microscopy,"* Science, 2012, pp.1325-1329, DOI: 10.1126/science.1225621

32. Meyer, J.C., Kisielowski, C., Erni, R., Rossell, M.D., Crommie, M.F., & Zettl, A., TEAM 0.5, Berkeley National Laboratory and Berkeley Lab's National Center for Electron Microscopy (NCEM), *"Direct Imaging of Lattice Atoms and Topological Defects in Graphene Membranes,"* 2008, Nano Letters, 8 11, 3582-6, https://pubmed.ncbi.nlm.nih.gov/18563938/

33. YouTube, @Synergetics University, *"The Synergetic Geometry of Buckminster Fuller | INTRO,"* https://www.youtube.com/watch?v=VYVUbxwqhyY

34. Fuller, R. Buckminster, Applewhite, E.J., Macmillan, *"Synergetics: Explorations in the Geometry of Thinking,"* 1975, 1979, 794.06, pg. 200, 794.06, ISBN: 002541870X

35. Ibid, pg.201, 795.08

36. Haramein, Nassim, *"Black Whole,"* Gaia TV, (Film), 2011

37. YouTube, @Luke Ragle, *"Vortex Based Mathematics Vector Equilibrium (Cuboctahedron),"* https://www.youtube.com/watch?v=8HcE2lIiumE

38. Grant, Robert E., *"Predicting Prime Numbers,"* 2018, https://archive.org/details/predictprime

39. YouTube, @Resonance Science Foundation, *"Resonance Talks • 'ONE Is The Only Constant' with Robert Edward Grant,"* https://www.youtube.com/watch?v=IZEAkRFTTbk

40. Grant, Robert E., Ghannam, Talal *"Accurate and Infinite Prime Prediction from Novel Quasi-Prime Analytical Methodology,"* 2019, https://doi.org/10.48550/arXiv.1903.08570

41. YouTube, @Resonance Science Foundation, *"Resonance Talks • 'ONE Is The Only Constant' with Robert Edward Grant,"* https://www.youtube.com/watch?v=IZEAkRFTTbk

42. Ibid.

43. Grant, Robert E., *"Are ALL Mathematical and Physical Constants Simply Mirrored Interchangeable Transformations of The Golden Angle?"* 2022, https://robertedwardgrant.com/are-all-mathematical-and-physical-constants-simply-mirrored-interchangeable-transformations-of-the-golden-angle/

44. Feynman, Richard P., Princeton University Press, *"QED: The Strange Theory of Light and Matter,"* 2006, ISBN 10: 0691125759 / ISBN 13: 9780691125756

45. Grant, Robert E, *" 'Precise' Temperament Tuning,"* 2020, https://robertedwardgrant.com/wp-content/uploads/2024/01/Precise-Temperament-Tuning-432hz-Robert-Edward-Grant.pdf

46. Bendall, Malcolm, *"The 'Thunderstorm Generator' - Atomic Energy From Water Plasmoid Protium Power,"* https://www.strikefoundation.earth/open-source-research

47. Bendall, Malcolm, Ascension Dynamics, *"Plasmoid Unification Model Overview - Malcolm Bendall,"* 2022, https://www.ascensiondynamics.org/blog-noteworthy/plasmoid-unification-model

48. Collin, Jordan, YouTube, @Alchemical Science, https://www.youtube.com/@AlchemicalScience

49. Bendall, Malcolm, *"Patent Application for MSAART EVO Technologies That Solve the Real World Problems of CO2, Energy, Power, Oxygen and Pollution,"* 2022, https://www.strikefoundation.earth/open-source-research

50. Elkins, Karen, Science to Sage Magazine, *"VBM: Vortex Based Mathematics with Marko Rodin,"* 2021, https://sciencetosagemagazine.com/vbm-vortex-based-mathematics-with-marko-rodin/

51. YouTube, @GlobalBEM, *"The Rodin Coil Over Unity Transducer | Marko Rodin | #VortexBasedMathematics,"* https://www.youtube.com/watch?v=XUTfdXs6QWY

52. YouTube, @AlchemicalScience, *"Malcolm Bendall's Plasmoid Unification Frequencies to Coherent Geometry - Sheela Rahman,"* Cosmic Summit 2024, https://www.youtube.com/watch?v=JsGz8G_VQJM; YouTube, @XAETHERYA, https://www.youtube.com/@XAETHERYA

Acknowledgements

I would like to acknowledge and thank the following for their assistance, knowledge and/or inspiration in the creation of this book.

Marko Rodin for the rediscovery of the inherent mathematical language of Creation, known now as vortex based mathematics. And also for sharing information, taking the time to meet with me about the book and connecting me with good people.

Lisa Wilson and Karen Elkins for their information, support and guidance with publishing the book.

Catherine Ensign for helping me create the book and for doing such a wonderful job on the editing.

Renee Giovando for the cover art design and creation, and for formatting.

Randy Powell and Greg Volk for their knowledge and insight.

John Snyder for analysis and feedback on the diagrams, and also for many hours of discussion regarding this knowledge.

Further thanks to Mike Trefrei, Steve A.A. Bauer, Jamie Buturff, Jason Collin, Randall Carlson, Luke Ragle, Sheela Rahman, Robert Edward Grant, Bob Greenyer, Nassim Haramein, Dan Winter, Malcolm Bendall, Charlie Ziese, Andre Nugess, Carl P. Munck, Patrick Flanagan, Wilhelm Reich, Walter Russell, Nikola Tesla and everyone else mentioned in the book. There are so many great teachers.

www.ingramcontent.com/pod-product-compliance
Lightning Source LLC
Chambersburg PA
CBHW080837230426
43665CB00021B/2873